Servicing Satellite TV Equipment

Servicing Satellite TV Equipment

Nick Beer LCGI, MIPRE

Newnes
An imprint of Butterworth-Heinemann
Linacre House, Jordan Hill, Oxford OX2 8DP
225 Wildwood Avenue, Woburn, MA01801–2041
A division of Reed Educational and Professional Publishing Ltd

A member of the Reed Elsevier plc group

OXFORD   BOSTON   JOHANNESBURG
MELBOURNE   NEW DELHI   SINGAPORE

First published 1998

British Library Cataloguing in Publication Data
A catalogue record for this book is available from the British Library.

ISBN 0 7506 3425 1

Library of Congress Cataloguing in Publication Data
A catalogue record for this book is available from the Library of Congress.

Typeset by David Gregson Associates, Beccles, Suffolk
Transferred to Digital Printing 2006

CONTENTS

PREFACE

The satellite television market is a phenomenally complex one from all perspectives. The overall perception of the medium is of low grade, expensive to view, often imported programming received by virtue of ugly antennae. The advent of populus satellite television in the UK primarily by virtue of News International's (Sky) BSkyB introduced us to a whole new concept of what format our TV broadcasts take and it has been decidely unpalletable to many. It has had a decidedly American taste and by no means the best of American taste!

It is a fact, however, that satellite transmissions to the UK, and indeed Europe, have led to a massive increase in choice of programming and some of this increase has been of very high quality. Possibly the most notable area of programming to benefit is that of sport.

One of the results of the extremely aggressive marketing of the programme providers – notably BSkyB – is that the hardware is often offered at ludicrously cheap prices – subsidised by subscription charges – often with free installation. The latter is often done by poorly trained and incompetent staff leading to conflicts with customers and genuine dealers. One has to understand the structure of the business to understand the non-technical constraints within which service engineers must work.

Big business saw the demise of BSB (British Satellite Broadcasting) many years ago. This was supposed to be the UK's official DBS (Direct Broadcast by Satellite) service but it was rather poorly conceived and run, and was swallowed up by Sky to form BSkyB, the current Astra based programme provider. What we lost here was the chance to have higher quality D-MAC (as opposed to PAL) transmissions; arguments persist as to whether this was a real loss!

Aside from the 'Sky' market, there is a small band of satellite viewers who use more specialist equipment – cost being realistic rather than subsidised – to view often very weak signals from multiple satellites. Also, they use more refined electronics in their LNB and receiver. This is the longer running market supported on the whole by specialist, low volume satellite receiver manufacturers. However, the mass market, around since 1989 in the UK, has seen satellite equipment produced by or for virtually all traditional major brown goods manufacturers. I say 'or for' as the UK market (and indeed much wider) is currently dominated by one major receiver manufacturer – Pace Microtechnology. This company produces receivers that are then badged for many brands. One doesn't really begrudge them this distinction – a well run, successful British company that has grown phenomenally in the last few years, and from an engineering point of view, one couldn't wish for much more from them – good support and generally good products.

This market position will be reflected by the large-scale inclusion throughout this book of the company's example circuits, block diagrams and photos. Whilst more obscure manufacturers' circuits can be interesting, the reader will want to be able to associate the information in the text with products that he regularly encounters. As I stated in my other title, *Servicing Audio and Hi-fi Equipment*, the inclusion of specific circuits does not imply unreliability. Any faults that I suggest are to aid explanation in the text and are not indicative of particular problem areas.

Nick Beer
Northam, Devon 1998

ACKNOWLEDGEMENTS

I would like to thank the following for their assistance in granting permission to use and providing figures for the book.

Claire Hunter, Pace Microtechnology, UK
Melanie Dickie, Eutelsat, Paris, France
Bill Collins, Astra Marketing, UK

Also, thanks to Teleste Cablevision, Hameg UK, Metrix UK, Lenson Heath/Triax.

1

INTRODUCTION

It is likely that in a workshop environment, and on field service, equipment other than satellite receivers is being repaired. However, this being the case or not, the following refers to general topics of servicing life. Certain points throughout the book will bear significant resemblance to points made in my other title in this series, *Servicing Audio and Hi-fi Equipment*. This is because my approach and philosophy do not change just because the type of equipment has!

Service necessities

A number of prerequisites are necessary to ensure successful service of any domestic electronic equipment.

Firstly, the operative must be fully conversant with both theory and practice. To attempt a repair for which you are not competent is both irresponsible, dangerous and usually unsuccessful.

Secondly, a suitable range of aids from chemicals to tools to test equipment is needed. Chapter 2 covers types and usage, but again without the knowledge of how to use them effectively they are useless. Another relevant point is that however nice it may look on the shelf a piece of test gear must earn its keep: don't be tempted to buy for the sake of it. It is folly to try to carry out alignment without the correct test and measurement equipment. In turn, slight misalignment caused in this way can lead to non-existent faults being chased.

Next, a suitable area in which to work. Workshop conditions are covered later in this chapter.

A number of requirements combined under the term 'back-up' include supply of service manuals, technical advice and spares. These are covered in detail later. Loosely connected with this is the preference to have experience of the equipment on which you have to work. For many, though, this is an unaffordable luxury!

Workshop conditions

Some obvious points perhaps but we need isolated benches, fed via transformers for work on live power supplies. Distribution of u.h.f. as well as satellite i.f. signals via isolated outlets – a simple system such as those discussed in Chapter 5 perhaps. If you work with positioners/motorised systems, then a motorised dish nearby with feeds to workbenches would be good. Normally a dish feeding signals to benches would be fixed on Astra but it may be useful to have feeds from say one of the Eutelsat craft also to allow checks to be made on various audio bandwidths, carriers and de-emphasis standards, not to mention different signal strengths. A monitor to view results on and at least one on which to soak test receivers are necessary. Ensure that geometry is correctly set, otherwise faults may not be noted.

It is important that the working conditions are suitable. The area should be well and evenly lit overall with sufficient extra (close-up) lighting, ideally with a magnifier. The area should also be well heated in the winter and well ventilated in the summer. It is also advisable to be able to ventilate in the winter without draughts: with the heating on full the workshop can become very stuffy, leading to tiredness and headaches – as if we don't get enough!

It is often not possible, but ideally one should be working away from a constantly ringing phone, intercom, etc., and such through traffic as field engineers, installers, shop staff and, most of all, customers. Carefully check on your insurance policies and make all of your staff aware of who should be allowed where and how they are covered should an accident occur. It is always better to be safe than sorry even so. If the workshop is isolated from the retail shop a phone or intercom is usually necessary, but it should be made clear that it should only be used for essential communication.

An effective computerised administration system should allow all job enquiries to be dealt with without disturbing engineers. The information is updated as the job is progressed by them. Anyone can then obtain status information. Without a computer system, a service administrator should perform the role of liaison between engineers and customers.

Field operations

Naturally, to operate in this field of servicing you will need to be able to reach dishes. Therefore steps and an extension ladder will be necessary and perhaps also a roof ladder. However, it is a fact that very few dishes are mounted above the eaves for planning and practical safety reasons. Consequently, ladder racks for your vehicle(s) will be required. There is now legislation requiring approved, minimum standard racks with adequate mounting on the vehicle. Do ensure that racks and mountings meet the required standard for the weight of your ladders. Ladder types are discussed in Chapter 2.

A vehicle kit will contain specific spares for the types of equipment that you encounter. However, there are some obvious generalities. LNBs of the Marconi type with 10 GHz and 9.75 GHz local oscillators. Even good second-hand ones are useful for testing and fault finding – never be without one.

Similarly, a test receiver. As long as it works, it doesn't really matter what it is. It is, however, invaluable in proving a fault quickly in the field. When you have sparklies and your meter test is inconclusive, popping another receiver on the signal will determine where the problem lies before you go getting ladders out and moving dishes! It is therefore important that your test receiver is regularly tested and that you know what level of performance to expect from it under specific conditions, i.e. Astra (UK Gold being a weaker signal) from a 60 cm dish and from an 80 cm. Meaningful comparisons can then be made.

Interconnecting r.f. leads and scart leads can be used to solve problems and to make money – try to sell them to customers! Obviously you should also carry a drum of suitable coax cable and some mains flex.

From a customer perception point of view, keep vehicles clean and tidy – help to avoid cultivating a cowboy image.

The fault

A vital point in the repair process is right at the start when the customer reports the symptoms, particularly if they are intermittent. There is a skill in dealing with the public and all staff coming into contact with customers via field service, telephone or shop should be clear on how to get the required information. What is more infuriating than getting a unit on the bench, or calling and checking all functions, finding no obvious fault and turning to the fault report for enlightenment only to find 'Repair as necessary' and – as if to rub salt into the wound – 'Estimate if over £5'!?

Always carefully question the customer. If he wants 'a service' find out if there were any fault symptoms that provoked this desire – there usually are.

Determine if symptoms occur on all or specific channels. If the symptom is elusive ask the customer to make a video recording of the problem and bring it to you – this can help diagnosis no end.

Communication

As good communication is expected between the customer and ourselves we must ensure that this is reciprocated. If a unit comes in with a particular fault which when rectified reveals other problems or potential problems, contact the customer and explain. Many customers require estimates for repair, which in themselves present problems (see later) but a two-part estimate to rectify reported and also other potential faults is often appreciated.

The repair

Having all the information from the customer is a good start but it must be complemented by a continuation of good practice. Before diving into a repair look at or listen to all the symptoms – there is commonly more than one and by detecting all the signs you can quickly narrow down the areas for investigation. This important aspect of diagnosis will be mentioned several times throughout the book.

Following this get the service manual wherever possible and read any notes or technical bulletins therein. This can save hours in both finding the fault and dismantling the unit for repair. When you've got the unit open inspect it closely. Look for signs of overheating, explosion, corrosion, spillage and arcing. All or any of these found will save fruitless checking. There will be some natural signs of heat in power areas so do not confuse this with a fault.

Likely problems

There is a definite pattern of component faults and bearing this is mind will often lead you to the fault in a circuit where in theory any component could be at fault.

These suspects should include:

Mechanical faults before electronic ones;
High value resistors of say over 100 KΩ;
Electrolytic capacitors in older equipment or areas where heat is produced: the capacitors tend to dry out;
Dry joints and printed circuit breaks where heat and heavy components are involved;
Controls, switches and sensors which involve mechanics; also plugs and sockets, all of which can wear or corrode; and anything external to the equipment such as batteries and aerials.

Intermittent faults

Modern electronic production and construction methods tend to give rise to faults of an intermittent nature. Many reasons could be suggested for this; finer and thinner print, thinner PC boards, smaller component legs, wave-soldering techniques and smaller unit enclosures, for example. That means that life for the service engineer gets more difficult – if a fault is intermittent it usually takes a long time to find, and the corrective action may not easily be proven to be effective.

To help overcome this latter problem it is always advisable, wherever practical, to try to simulate or provoke the fault. There are two approaches to this. Firstly, the symptoms described may suggest one or maybe a couple of likely faults to you, whether this be based on previous experience or from information gained

from a book like this. If you do suspect a particular fault try to simulate it safely. Lifting components to simulate open circuits can be fruitful but do ensure you know what you are doing. Use these techniques where there is no real power involved.

The philosophy is that it is always better to have some proof, however circumstantial. You will always feel a lot happier if you know that the part you have changed on speculation could actually cause those exact symptoms.

The other approach to intermittent faults is used when you have another unit to hand of the same type in which you can swap the suspect part in order to prove the point. Transplanting the fault thus provides double assurance.

Adjustments

Reference to alignment and adjustment will be made in a general sense throughout the book. Individual service manuals should be consulted when performing a specific adjustment and the specified test equipment should be used.

It is tempting to 'twiddle' adjustors in a fault condition to see if and how they affect the fault – and indeed to see what the adjustment does in the absence of any service information. This can be very useful but also very dangerous. Always write down what you do, recording the original settings and when you have finished reset the controls to specification as per the service manual. Before you start, then, you should ensure that you have the means to do this, most satellite receivers (at least mainstream) have few internal adjustments.

Repair estimates

It is a fact of modern life that most customers require estimates. This makes the job longer, more difficult and more expensive. Estimates are required for two main reasons: normal viability concerns, and insurance claims where accidental (and sometimes not!) damage has occurred.

To cover the costs of preparing estimates it would appear to make sense to charge for those refused, but this is seldom practical especially as many companies advertise 'free estimates'. There is also, then, a tendency for owners not to collect units upon which they have refused estimates. It is reasonable to charge for supplying a written estimate though, and having to deal with queries from the insurance company or loss adjuster.

When estimating for a normal repair, always allow a little leeway in the price if you have been unable to prove the fault or if you cannot fully test the unit. Ideally you will do enough to get the unit to a state where all other functions can be tested. It is extremely unwise to give a firm price for a dead receiver repair if you haven't got it working and tested it – there may be all kinds of other problems.

Try to get an immediate decision from the customer. All over the world there are thousands of uncollected, unrepaired appliances taking up a lot of space. Find out if the customer wants the unrepaired unit back or whether you can throw it away. If it is required always reassemble it as far as possible into the condition in which you received it. This saves a lot of potential problems.

Equipment lifespan

Receiver lifespan is directly affected by cost of repair versus cost of replacement. All consumer electronics products have come down in price year on year in real terms whilst, naturally, repair costs have been affected by inflation. The satellite market is even more peculiar due to the subsidised cost of the hardware made available by the service providers, such as BSkyB, to entice subscriptions. Furthermore such schemes as a free repair service have been offered here in the UK to prevent loss of subscribers. All this paints a fairly bleak picture for repairs which is in fact not a true reflection. There is plenty of work about and people will (eventually) pay for it. The cheap deals that they immediately think of when they have to face a repair bill involve lower quality equipment installed by less competent companies and are no replacement for their faulty unit.

Early receivers suffered badly with poorly designed power supplies (not always linear) that overheated, resulting in regular failures and damaged PCBs. The life of a mainstream receiver is likely to be little more than five years, especially with the emergence of new technology such as digital services. Specialist receivers, built to a design rather than a price, are on the whole very heavily built and their lifespan is likely to far exceed this five years.

Outside kit – the dish, etc. – may last less than a couple of years in a rough or salty climate – early dishes rusted badly in any climate. Fibreglass dishes in a sheltered area will last a very long time. Some manufacturers now give a ten-year guarantee on their dish reflectors. No simple answer there then!

Service information

Depending on the type of service organisation concerned, the matter of service information and technical training will vary from readily available and commonplace to rarely available and an unknown luxury. Ideally, we would like to work on a limited range of equipment with good spares and service information back-up from the manufacturers concerned, and be familiar with most products encountered by attending regular training courses and having in-workshop training. If you are a 'one man band' engineer working on his own who has to accept virtually anything for repair for financial reasons, you won't have these benefits, which make a tremendous difference to workshop efficiency.

It is to the author's benefit that he currently works such a system. Makes are limited to six or seven, service manuals are in most cases directly mailed on release, and senior engineers attend regular seminars. The information acquired is then passed on to other engineers at 'in-house' training sessions. In many workshops such training is largely neglected due to pressure of workload. However, this type of situation should be striven for.

Repairs to mainstream satellite receivers are slightly different in this respect to the advantage of those doing them. There may be many different brands, but few manufacturers. This means that a single circuit may suffice for many makes and models. Furthermore, satellite equipment tends to be distributed through wholesalers not by manufacturers, and these wholesalers tend to have their own repair facilities making it possible for a dealer with no service capability to sell the product; similarly, aerial contractors with no electronics ability. Pace are, of course, the major badge manufacturer and they make training available to anyone within the industry – similarly their technical advice service and repair facilities. It is far easier therefore for someone to become involved in repairing their products than tradition is in the brown goods field. You still need to be competent though!

Manufacturers' technical advice services vary tremendously in their usefulness. If you are stuck for a lead and consultation with fellow engineers proves fruitless (or you have no fellow engineers!) a call is undoubtedly worthwhile but the author's experience has been as follows.

If the fault is well known by engineers throughout the trade who are experienced on the unit then it will be known to the manufacturer. Therefore if you aren't familiar with the equipment there's a very good chance that they can help. If the equipment is fairly new and the fault is difficult to diagnose or cure there could well be a modification or production error and this is where manufacturers are invaluable. The modification you need is always the one that was on the technical bulletin which got lost or thrown away or which, indeed, hasn't yet been produced! If you are familiar with the equipment and cannot find the fault then in my experience the manufacturer's only help will be to describe the operation of the circuit, which may be the key to diagnosis of the fault but all too often is not. How many times have you telephoned a helpline only to receive the stock reply: 'We've never had that one before, Sir/Madam, do let us know if you find out what it is!'?

No-one is to blame for this of course. Manufacturers can only help if they've encountered the problem, otherwise it is down to informed speculation. Many provide a sterling service.

Spare-part ordering

One of the most time-consuming and frustrating parts of the repair process is obtaining replacement parts. Some basic guidelines, then, may help to make the process as painless as possible.

Wherever possible use the service manual to identify parts and obtain the manufacturer's part number. Hopefully, all corrections sent from the manufacturer are updated, by you, in your manuals. Double-check the description and part number that you copy down and be particularly careful not to confuse parts when using exploded diagrams.

When using a manual that covers units of different models, colour/finish, etc., and those for other countries, make absolutely sure that you have the correct part number for your model, colour and country where applicable.

Another benefit of having few manufacturers of satellite equipment is that spares are available from a majority of sources under different brand names. A few parts will not be interchangeable – cabinet parts with brand names on them, for example.

As mentioned elsewhere, if a component is designed as a critical safety component by the marks $\diamondsuit$ or $\triangle$ use an original approved replacement.

An eye should be kept on long outstanding spares orders. If you've ordered three or four parts and one hasn't arrived, try fitting those you have plus an equivalent part, where permissible. If you are hoping to cure a fault by fitting an IC why wait an extra three weeks for a knob to arrive as well, only to have the diagnosis proved wrong? Better to start the repair as soon as possible.

With regard to service manuals and technical bulletins, it is vital to store all information received in a logical order and in way that preserves their life. No matter how simple a repair or adjustment may seem always get the manual out first. You may repair a primary fault only to find it bounce back on you. If you'd checked with the manual you would have seen the note you put on it some years earlier or the mod-sheet about checking another part as well.

Never be tempted to throw away a manual, as for sure that model will come in soon – like the day you threw out all those old valves and then realised someone would have paid you a small fortune for them.

The other important aspects of workshop life are covered in the next chapter and in Appendix 2; addresses for spare parts are given in Appendix 3.

2

TOOLS AND TEST EQUIPMENT

The precise collection of tools and test equipment will, of course, depend on the extent of your sphere of activity. If you do not repair outside kit, dishes, etc., there is little likelihood that you will need spanners and ladders! Conversely, if repairing receivers, it is very likely that you will also be working on other consumer electronics such as TV and audio and so be using much common equipment.

Hand tools

Musts are a selection of screwdrivers from the smallest jeweller's sizes up to heavy duty, long shaft versions (e.g. no. 3 cross-heads). Trimming tools (insulated) are also necessary. Pliers from fine long nosed (and tweezers) to heavy duty combinations and mole grips. Spanners up to 25 mm (if polar mounts encountered), with more than one 10 mm, 11 mm, 12 mm and 13 mm as these are common az/el mount bolt sizes. Naturally adjustables are also valuable. For bolting dishes to walls (even repairs rather than installation calls for this intermittently) a socket set and torque wrench are needed. The latter to ensure that wall bolts are not over tightened – doing so can lead to structural damage.

Ladders

Legislation requires that ladders are of a certain standard. Tradesman grade is required by most manufacturers' definitions and are typically double extension. Such ladders will be heavier than the sort you have for DIY around the home, and working all day with them needs to be considered when purchasing. Very lightweight alloy ladders, triple extension for greater convenience, are available and make daily life much easier, but the cost of such items is rather more than a standard double extension. A roofing ladder, which

has wheels and a hook to enable it to be located over a ridge, is occasionally necessary but most dish installations are below eaves' height. If you don't have a roof ladder, you may well get caught out!

One cannot overemphasise the need to get the right ladders and to ensure that they are used and maintained safely. No-one should ever use ladders without being fully conversant with how to handle them properly. Ensure that you always have stays, ropes and eyes in order to lash ladders on site. Your vehicle must obviously be capable of carrying ladders securely and must have the requisite ladder racks meeting relevant safety rules. A red flag should always be attached to any ladder overhang on a vehicle.

Signal strength meter

As a service engineer there is no excuse for not using a self-powered, tuneable meter. Being able to power the LNB without a receiver connected is essential for fault finding work, as is being able to tune in the first i.f. band (see Fig. 2.1). The in-line sat-finder types may be fine for basic installers but not for diagnosis. A good meter will pay for itself time after time. There is no way you can contemplate life without a signal strength meter. Many such meters will incorporate u.h.f. and f.m. ranges which are invaluable when working with distribution systems or even r.f. spacing of domestic receivers. Other useful features are the ability to measure low d.c. voltages (LNB or masthead amplifier supplies), resistances (at least to determine continuity) and measure LNB current.

Field monitor

A battery powered, lightweight monitor – ideally colour – is very useful to provide definitive quality tests. This is especially relevant when working on

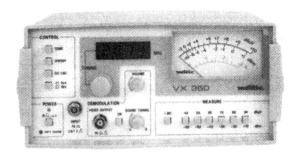

Figure 2.1   *An example of a good, general purpose signal strength meter suitable for installation but more especially suitable for servicing work. It is self-powered and has tuneable, readable i.f. (METRIX)*

a distribution system where you may want to assess many different points but not have access to individual properties' equipment. Many designs of spectrum analyser have these facilities built into the one unit.

Spectrum analyser

This piece of test equipment is automatically considered by many to be too expensive without them investigating what is available. As the satellite market has grown, suitable, more economically priced units have been developed and now

they are within the reach of anyone doing any volume of work. They, like SSMs, will often include u.h.f. and f.m. ranges and so their use is widened.

The idea is that you view a graphical trace (not unlike an oscilloscope display) of signal amplitude across a given bandwidth of frequencies (Fig. 2.2). This allows a spectrum – e.g. the first i.f. from an LNB to be viewed and relative amplitudes of channels or vision to sound to be examined. Frequency related faults – a mismatch, for example – can then be clearly seen by their effect on amplitude around the given frequency.

At their simplest, this is a spectrum analyser. The more you pay, the greater the accuracy and resolution you will return. Extra features on even modestly priced units include a monochrome monitor to view off air signals, back porch analysis which displays an oscillogram of the sync and burst area prior to the vision. (See Fig. 2.3.) Again problems with reflections or text corruption can be highlighted here. More advanced features include teletext decoders, tuneable audio demodulators and Mac decoders.

Test equipment for digital signals

Equivalent and compatible equipment for the rather specific demands of digital signal measuring

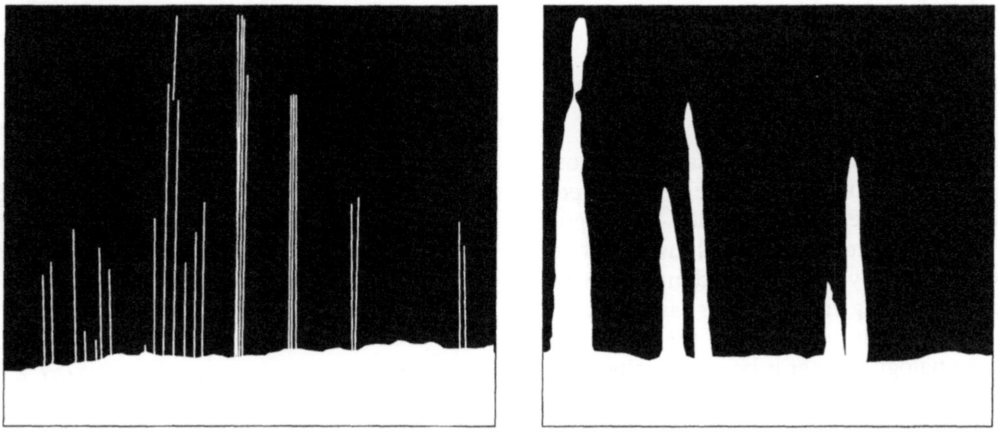

Full band	Expanded area

Figure 2.2   *A spectral display as produced by a spectrum analyser. One can view the vision and sound carriers in the displayed bandwidth, their relative amplitudes and bandwidths. The resolution is usually switchable to enable a specific area of the band to be magnified.*

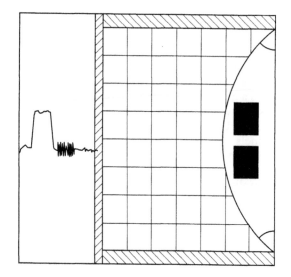

Figure 2.3 *A feature often employed on spectrum analysers is the ability to view the line sync pulse and burst (back porch) on screen prior to the vision*

and monitoring are now becoming available. It's best to wait and see what your requirements will be for this before investing lots of money! Most digital signal measurement equipment analyses the data-stream but presents its 'verdict' in a manner analogous to an analogue system, i.e. equating picture quality to the display and bit error rate.

Multimeter

A plethora of meters are today available and cost varies from a few pounds to several hundred. Digital multimeters are the only choice – analogue is simply not adequate for general use, and in many cases can lead you a merry dance with its low impedance loading circuits. This said, a digital meter with bar graph to simulate the action of needle on an analogue meter is best – the slowness of digital displays means that you cannot get a 'feel' for a test as you can with a needle. Checking the charge of an electrolytic capacitor for example.

You will need the ability to measure a.c. and d.c. volts up to 1000 V, current up to a few amperes and resistance to 20 MΩ. A diode test

facility is also very necessary. Go for portable meter with 3.5 digits minimum and 0.1% accuracy and you will be OK. Spend too much and you'll be afraid to take it up onto a roof! If you have the budget, then a portable DMM (digital multimeter) for field work could be complemented by a larger bench-mounted unit in the workshop and an analogue meter for its limited component testing uses.

Oscilloscope

Use in the field is likely to be limited and so considerations of portability are minimal. A bench scope of say 40 MHz bandwidth will suffice for satellite receiver work but in most workshops this will be in company with repairs to other equipment and so the specification may be higher. Lower bandwidth 'scopes will suffice for much work but their features tend to be more limited and the displays are dimmer. This said, the use for delayed sweep and the like is minimal. Preset triggering for TV field and line are highly desirable (see Fig. 2.4).

The 'scope is one of the most useful tools an engineer can have. Many feel apprehensive about using it more than they have to, while others use it

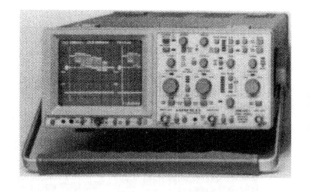

Figure 2.4 *A modern oscilloscope is ideal for general workshop use. This incorporates new, electronic switching, on-screen displays enabling much greater accuracy of measurement and a digital storage facility. Units with less features are perfectly reasonable for receiver repairs and alignment but this specification enables faster, more detailed fault finding to be undertaken (HAMEG)*

for everything from signal continuity checking to voltage measurements and alignment. To use a 'scope to look for a fault on a portable radio may be regarded as 'overkill' but used properly and efficiently it makes the job much easier and quicker – so why not? D.c. voltages can also be readily checked and if the 'scope is up and running it can be easily used for this purpose by simply d.c. coupling the Y input. This saves changing to a meter if simple tests are required.

Digital storage oscilloscopes

Digital storage in an oscilloscope can be very useful. It allows a waveform to be held in memory for an indefinite period of time (in some cases, as long as the unit is powered) thus making the inspection of very slow waveforms or transient phenomena very easy. Other advantages of storage 'scopes include the ability to 'record' a single-shot waveform, facilitating a 'baby-sitting' role. Where, for example, a supply rail is suspected of having intermittent noise the DSO can be set to trigger on the noise pulse and left running until it occurs. The engineer can then go back at a later time and view the stored waveform. The cost of DSOs is coming down (DSOs with a 20 MHz/20 MHz specification can now be sourced for well under £1000 in the UK) but the cheaper units, whilst having an adequate analogue bandwidth, have a restrictive digital sample rate.

Newer DSOs will allow interfacing with PCS to store waveforms but also the PC to control the 'scope. This can be especially useful in development work.

Probes

An oscilloscope probe is a fairly delicate device and should be treated as such. Replacements are readily available from suppliers but are relatively expensive. The use of modular probes is sensible because if leads or tips get damaged each component of the probe can be replaced separately. Each probe includes a trimmer to allow it to be aligned for the input of characteristics of any 'scope. Fig. 2.5 shows the correct procedure for this, using a square wave reference output from the 'scope.

It is advisable to use 10:1 probes to prevent loading of the circuit under test. Many probes are

switchable between 1:1 and 10:1. In the event of a faulty reading the probe should be checked: it could be that it is switched to 10:1 and the user is not aware of that; alternatively it could be set to 1:1 and thus loading the circuit excessively – switch it up to 10:1 and try again with the appropriate resetting of the Y amplitude.

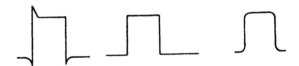

Figure 2.5 *Optimum and incorrect oscilloscope probe alignment: the centre waveform is optimum and the outer two should be avoided. The probe is connected to a square wave source, usually provided as a test point on the front of the oscilloscope.*

Signal generators

Most workshops will contain an audio generator. When you are repairing audio faults, it will prove useful. However, it is difficult to say that you will *need* one to successfully repair satellite receivers though. An r.f. or i.f. generator capable of working in the audio i.f. stages may be similarly useful but the same reasoning applies. Any thoughts of a microwave generator for LNB testing is likely to be fanciful due to their extreme cost and limited if existent necessity. (See also Chapter 4, LNB repairs.)

Bench power supply

A d.c. bench supply capable of delivering 0–30 V is invaluable. It will be used to inject external supplies for test purposes – LNB supplies or driving actuator arms, for example. It should have up to 30 V so that we can replace the tuning supply (BT). Various techniques are mentioned throughout the book. Current capability for satellite service need not extend beyond a few amperes.

Frequency counter

A reasonably accurate frequency counter is essential for setting up of oscillators – chroma, clock, PLL, VCO, etc. Those found on DMMs are not usually sufficiently accurate or sensitive. A bench unit is not expensive.

Portable appliance tester

This relatively new piece of test equipment is necessary to ensure that repaired equipment is safe. Its ease of operation and complicity with UK regulations makes the traditional insulation tester redundant. The use and detail of portable appliance testers is covered in Appendix 2.

Infra-red output detector

With remote-control handsets in proliferation some means of checking them for emission is required. Several testers are available, usually acknowledging output by a flashing LED or audible tone. The cheapest way of checking is to connect an i.r. receiver diode to an oscilloscope and check for a waveform, though this can be a little inconvenient! A recent development is the i.r. *mirror* type device in the form of a card whose surface reflects an orange glow when i.r. light hits its active area. A spatula-shaped variant is available which lends itself to checks of CD laser power as well as emissions from photo-interrupter type devices. A rough check on i.r. handset emission can be made by 'squirting' it at an a.m. radio tuned to medium wave. Many camcorders are sensitive to i.r. radiation and if a working remote handset is pointed at the lens and operated a light 'splodge' will be seen on a monitor connected to the camcorder.

Infra-red handset tester

A unit that will give an audible or visual indication (or both) when i.r. light is emitted from handset pointed at it. A simulated range test is also sometimes included. The test should be taken in context – it only confirms presence of light from the handset, not that the modulation is correct.

3

DISHES AND FEEDS

The most visible part of the overall satellite signal reception package is the dish and its associated mounting, which are also the least technical. There are, however, some less than obvious problems that can occur and we shall see this after dealing with the basics.

The dish reflector

The parabolic shape of the dish antenna is the heart of its operation. The simple equation is that the more accurate and undistorted the parabola, the more efficient the dish. In practical terms, the most likely time to encounter a distorted dish is upon installation – twists or dents caused by the transportation process. It is a fact that as greater numbers of dishes are being mass produced, the mechanical strength seems to be becoming less, due to the use of thinner materials, and the standard of box and packing is also dropping. Often arms and brackets for the kit for the dish are found loose inside the box and small marks are evident on both sides of the dish. This is, of course, all symptomatic of a high volume and acutely cost conscious market.

It is therefore wise to check dishes at least before you mount them, or, if travelling a long way before you leave base, to ensure their integrity. Stand the dish upright on the ground and take a sideways look at it to spot twists, alternatively use a straight edge from rim to rim.

Incorrect assembly: over-tightening bolts can cause dishes to buckle. Good manufacturers will go to the trouble of advising maximum torque to be employed.

These comments have pretty much assumed that we are using metal dishes, which are the norm in mass market satellite reception. Let's now consider all different parabolic reflector constructions.

Reflector constructions

Sticking with metal for the moment, there are two pairs of classifications to make about the reflector. Firstly is it solid or perforated/mesh, and, secondly, is it steel or aluminium?

Generally speaking, a steel dish is more prone to oxidising (rusting) than an aluminium one, the latter being the slightly more expensive. Many older steel solid dishes were pressed and these can rust badly leading to a layered expansion as the rust appears blistering from under the painted surface.

Non-solid dishes can be either perforated or mesh. The difference being that perforated dishes' holes are larger but fewer and with more metal between them – a bit like a colander (see Fig. 3.2) whereas a mesh dish has many fine holes, very close together. The performance against solid dishes is inferior but to an extent that is not too significant with Astra transmissions when comparing 60–65 cm dishes. It is weighed against their perceived aesthetic benefits. This is all rather amusing as mesh dishes are invariably black and it is amazing how many people have had them installed on their white or light coloured homes – a solid, off-white dish would have been significantly less visually obtrusive, especially now as many early mesh dishes are rusting badly leaving long brown streaks down over the walls! Mesh dishes can be buckled more readily than solid ones – for instance, when being transported, installed or during heavy weather.

Non-metallic dishes

There are a number of dishes on the market that are made from composite or fibreglass type materials. Generally speaking they offer the advantage of not oxidising. Being lighter and less susceptible to damage also endear them, although when they do receive a knock, they can crack and

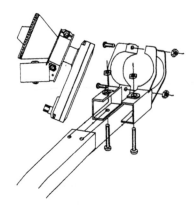

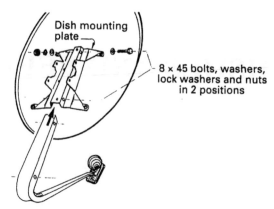

Position feed/LNB into LNB holder and secure with the 2–M5 × 14 screws and nuts. Push clamp and feed/LNB onto feed arm and secure with the 2–M5 × 30mm screws and nuts supplied.

Insert feed arm into position on the dish assembly and use the 8–45mm bolts provided to firmly secure.

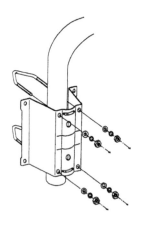

After fixing the wall plate to a suitable south facing wall (it is very important that the wall plate is fixed to the brickwork and not the mortar), position the all tube into the wall plate.
The outdoor unit must be fully tightened after elevation and azimuth settings are complete.

Figure 3.1 *Assembly of a popular, better quality dish typically used in fixed installations. This example is for an 80 cm reflector*

be useless. There have been odd problems with the material used not having good UV (ultraviolet) stability and the dish de-composing after some time in the elements. One down side of their use is their higher cost.

When is it necessary to replace a dish?

A question with no single answer! A rusty dish or one covered in flaking paint may look awful but will often not give rise to symptoms that the viewer perceives as annoying – witness the many such dishes around your area.

There is no question that dishes in this condition are not performing properly, as are those which are dented in storms and the like, but the only practical answer is to replace them when required to clear a fault that you are asked to clear.

A fair question might be how to determine that the rusty dish is causing poor signals – it could be the LNB, receiver or cable. How you decide will depend on your skill level and what test equipment and spares you have around you.

Later, we shall look at how to troubleshoot

systems but a point to bear in mind is a common-sense one – choose to replace the dish last!

Painting dishes

A much asked question is 'Can I paint my dish?' The answer is yes, but only using the correct materials. The answer to a rusty dish may be a rub down and repaint, but the cost of a new dish means it often makes more sense to replace it.

Others may wish to paint a dish to make it blend in rather better with its surroundings. Careful, even application of a non-lead based paint which is less than 30% reflective will suffice. If it is too reflective, heat reflection from the sun via the dish will cause problems with the LNB.

Dish antennae configurations

It is important to be able to recognise the different designs of dish so that you are able to align them properly. If you arrive after a storm to find a prime-focus dish off beam but you don't recognise it as such, you will not have much joy by trying to align it as you would an offset! Figure 3.4 details all the various dish configurations. 99% of fixed Astra type dishes will be of the offset type. The operation of the various dish types is well documented in other titles (see D. J. Stephenson, *Newnes Guide to Satellite TV*) by the same publisher. The rudimentary point that must be appreciated here is that whilst prime focus dishes look directly at the satellite and thus will have a relatively steep look angle, offset dishes do not, and will in most sites throughout northern Europe look almost upright, i.e. parallel to the wall or pole that they are mounted on. It is vital to have an appreciation of your site in terms of direction (azimuth) and look angle to be able to quickly and correctly determine whether the dish is clearly off beam.

Dish alignment

In the context of this book, alignment would be encountered only as part of a repair – typically where a dish has been blown off beam by wind – or where reflector replacement has been necessary. However, it is often the case that expert

Figure 3.2 *An example of a perforated dish. Here we see a 48 cm version with a universal LNB fitted ready for higher power (including digital) transmissions (TRIAX/ LENSON HEATH)*

Figure 3.3 *Your views invited! A motorised dish camouflaged by a plastic cover to look like a rock! (TRIAX/LENSON HEATH)*

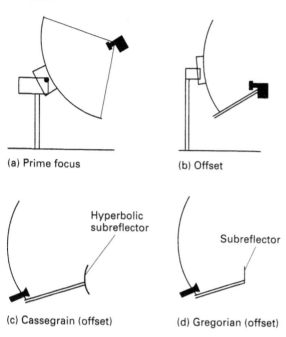

(a) Prime focus

(b) Offset

Hyperbolic
subreflector

Subreflector

(c) Cassegrain (offset)

(d) Gregorian (offset)

Figure 3.4 *An illustration of major dish configurations. Cassegrain and Gregorian can be configured as prime focus or offset as shown here. The vast majority of dishes used in Europe, fixed or motorised, are offset*

installers and service engineers are called immediately after installation by a third party due to problems being encountered. As stated in the Preface, rather less than expert installers are aplenty, having been given an intensive course (typically a couple of days) with no technical background and thus some truly remarkable fundamental mistakes are encountered. An often repeated scenario is that the customer becomes so exasperated by the incompetence that they willingly pay for someone competent to sort out the installation. Examples of dish alignment problems are many but one memorable case was where the Astra dish had been fitted looking north on a north facing wall!

It is important therefore to understand the prerequisites of installations, as many an external fault is caused (however much later in time) by something done at installation. More such examples follow in Chapter 5.

There is absolutely no alternative to aligning a dish with a signal strength meter. Yes, you can get signals without but you will *never* know if you are

spot-on beam and thus you cannot guarantee that signal levels will not vary intolerably with changing atmospheric conditions. Apart from anything else, how professional do you look, aligning a dish by watching the customer's TV through a window?!

You can start by setting the elevation of the dish. This can be achieved by your experience of the area or by looking at an appropriate chart. Set it by using an inclinometer on the rear plane of the dish (refer to dish manufacturer's instructions if you do not know where this is) and moving the dish to obtain the required reading. Tighten the elevation bolts so that they will not allow the dish to move. Then with your meter on a relatively low range gently move the dish towards the required azimuth – invariably from south towards the east (e.g. for Astra or most Eutelsat craft). As the signal increases and your meter attains full scale deflection, knock it up a range and carry on until you attain the peak signal. Lock the azimuth bolts and then ensure that your elevation setting is giving maximum signal. Having fine tuned both planes of adjustment, lock the bolts so that they are not going to allow the dish to move. When you tighten bolts, do so equally, i.e. one turn on the left hand bolt and then one turn on the right, all the time watching your meter to ensure that you do not pull it off beam. Do not over tighten and crush the pole on which the bracket clamps. This may render it impossible to adjust and retain a new position for the azimuth. It is a fact of life that however carefully you lock a dish, in certain situations they will move in the wind. A persistent problem may be answered by choosing a new site which is more sheltered, but if a site is on high exposed ground there is little anyone can do. In certain cases, welding the brackets to the pole having made doubly sure that alignment is correct has helped. The consequences of this though many be to remove a mechanical stress point that will then result in the dish being blown clean off the wall or bracket rather than just shifting.

Weak signals

This can obviously have many causes. Reference to the flowchart in Fig. 3.5 a and b will provide quite good guidance by simply substituting 'no signals' with 'low signals'. However, having decided that the problem is outside rather than

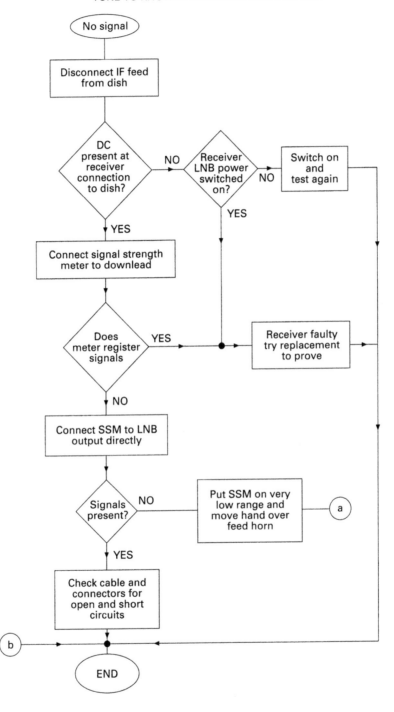

Figure 3.5a *Flowchart to enable diagnosis of where a fault lies in a fixed system*

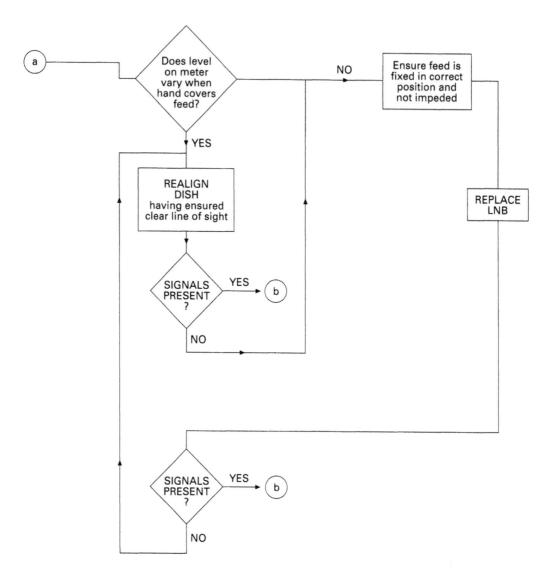

Figure 3.5b *Flowchart to enable diagnosis of where a fault lies in a fixed system*

with the receiver, we need to establish what may be causing the problem. Common sense will tell you that after some windy weather, the likely problem will be that the dish is off beam. Simply try realigning to see if you improve the level on your meter. It is important to remember that tiny inaccuracies in dish alignment – 1 degree, for example – not only reduces the level of signal from your desired satellite but also increases the possi-

bility of inter-ference from closely located adjacent ones. The role of the dish is thus dual: to maximise required signals and minimise adjacent satellites' ones.

A distinction that has to be drawn is that between weak signals that are a fault condition and those which are due to the basic inadequacy of the system. An old receiver with 60 cm dish and LNB with a noise figure of say 1.8 dB is not

Recommended dish size for optimal reception (ASTRA)

■ Individual dish 55–60 cm

▧ Individual dish 75–80 cm

☐ Individual dish 90 cm plus

Figure 3.6   *Recommended dish sizes for the UK for the Astra satellites (Astra)*

going to produce sparkle-free results from Astra, especially when there is heavy cloud cover. You may well find yourself in the position of upgrading a dish to a larger size one – similarly if you are replacing a damaged or rusted dish, you may try to upgrade your customer for their own benefit. Figure 3.6 gives the required dish sizes for Astra in the UK. It can be seen that here in North Devon, we are on the threshold between the 60 cm and 80 cm recommendation and there are many 60 cm dishes around. There is always pressure to use as small a dish as possible but you can't have your cake and eat it – and you have to tell people that! You also have to bear in mind that as satellites

age, their power may drop and so a system that was borderline at installation will only get worse.

The test is to peak the dish, try a replacement LNB (always carry one as part of your test kit) and see if the level on your meter improves. If not and assuming you have an unimpeded line of sight, then you have got optimal results with what you have.

Clear line of sight

When you are called to weak signals and the installation has been there for a while, it is often the case that the dish's view of the satellite is impeded. The usual cause is growth of vegetation – plants simply growing in size or deciduous trees developing foliage. Signal level variation is a good indicator of this as a problem – especially as the wind blows the plants around. Vegetation close to the dish and head end will also introduce noise to the signal, especially when wet. From behind the dish, look at the look angle – use a sighting compass if necessary and see what appears in your field of view. It should be a clear sky.

Booms and feed mounts

Problems here are few and simple. Bent feed arms/booms occur when the weather is inclement – often the only answer, especially on smaller dishes, is to replace the whole dish due to non-availability of parts. One regular failure is the clamp used to secure the feedhorn (or complete LNB) to the end of the dish boom arm. The favourite design is for a saddle clamp. This can be seen in Fig. 3.7. These are usually plastic and they crack. When this happens, they fail on one side and so LNBs do not drop out but become loose enough to swing in the wind causing signal level variations and problems with offset (see Chapter 4). Replacement clamps are generally available. Do ensure that the girth of the LNB is matched with that of the saddle clamp. Obviously too small a clamp will break again, and too large a clamp will result in the LNB moving.

Polar mounts

In much of what we have considered in this

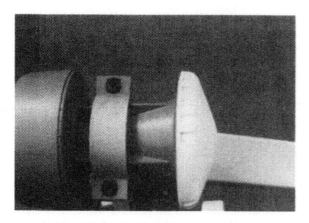

Figure 3.7 *Marconi type LNB showing saddle clamp and shoulder of LNB allowing for focal length adjustment*

chapter, we have concentrated on fixed dish installations. The huge majority of installations are fixed. Similarly, much of what has been discussed applies equally to fixed (Az El) and polar mounts. It is the truly competent satellite installer who becomes successfully involved with motorised systems – alignment is rather more involved. For a detailed description of the principles of operation and alignment of polar mounts, see *Newnes Guide to Satellite TV*. Here then is some practical servicing information. Keep in mind the interactions of all factors with a polar mount when looking for a problem. The usual scenario is that signals are fine from certain satellites but are poor or missing from others.

Tracking problems

The first thing to establish is that the problem is not one caused by incorrect positioning of the dish by the positioner unit. Chapter 6 considers this common problem where the electronics fail to return the dish to the correct azimuth position. Simply driving the dish either side of where it stops will determine whether this is the problem – if you find good signals by moving the dish position manually, then you have a problem with the electronics getting it right – this may ultimately be a mechanical problem with the mount but refer to Chapter 6. If you simply cannot get good signals or any at all despite being in the right area of the arc (azimuth), then you need to remove the actu-

ator (positioner motor) and attach the signal strength meter to the LNB. Move the dish through its arc by hand and monitor the signal level.

Why problems occur

The usual reason is movement or subsidence in the ground or storm damage or wind. It is vital to ensure before getting involved in realignment that the pole onto which the polar mount is fitted is perfectly upright – test with an inclinometer or spirit level in all planes. There is usually a degree of adjustment in upright stands to allow for resetting after movement or for installation on surfaces that are not level. Also ensure that all pole bolts are tight. Any missing bolts must be replaced and any tight or sticking areas cleaned and relubricated. Figure 3.8 shows the declination offset adjusting bolt of a popular polar mount, but above it can be seen where a rawl bolt has been used to replace the original fixing bolt on the actuator arm. The bolt is too small and so there is movement in the arm causing tracking problems.

Figure 3.8 *Declination offset adjustment also showing incorrectly fitted bolt top right*

Example

Let us say that signals from eastern satellites (Astra, Eutelsat 2F3 and 2F1/6) are fine but Eutelsat 2F2 is weaker than it should be and Intelsat 601 (at 27.5° west) is very poor bordering on non-existent. Metering the signals confirms this – allowing for the relative differences in power of course. This indicates that the polar mount is not facing true south as it should and thus cannot track the arc correctly. We now need to slacken off the elevation adjuster and determine, with the dish held at 27.5° west whether we need to increase or decrease the elevation to increase the signal level at this position. This will tell us whether the polar mount is too far east or west. If the elevation needs increasing, then the polar mount needs turning to the west so that it lands later, and conversely if it needs lowering of the elevation, then the polar mount needs turning to the east so that it lands earlier. Gentle and slight movements are what is required – try manual swings of the dish after each to confirm the effect. The elevation and declination offset (see Fig. 3.8) should be reset to the correct angle for the latitude of the site before commencing this. It is not necessary to compromise these settings to get results. These figures are cast in stone for each site and once set do not need moving again. Any problems left will be due to the pole not being upright or the polar mount not being accurately aligned to true south.

Feedhorns

Where, as with most fixed dish systems, a combined feedhorn, depolariser and LNB are used (referred to as a Marconi LNB), the feedhorn has little significance in the service sphere. One vastly overlooked fact, however, is that a feedhorn must be matched to the dish and boom to which it is attached. The area of the face of the feedhorn, the diameter, length and distance from the face of the dish are all important and have major effects on the impedance match of the interface between the two. Problems here can lead to a mismatch and problems with VSWR (voltage standing wave ratio) where cancellation could occur. In practice, as long as the correct feedhorn is used and the focal length aligned with a signal strength meter, there will be no problems. Even with the aforementioned 'Marconi' LNBs, the shoulder is broader than the saddle clamp into which it fits on the dish boom. The unit should thus be slid back and forth to obtain maximum signal level before tightening up the clamp. This arrangement can be seen in Fig. 3.7.

In more complex systems – motorised multi-satellite or older fixed systems – the feedhorn will often be a separate entity bolted to the depolariser which is in turn bolted to the LNB. Here is where there is the potential for a mismatch as many situations have been encountered where no thought has been given to matching, and whatever combination of the three units has been available has been bolted together. Where you are investigating a problem with such a system, never overlook the possibility that someone before you has made a mistake. Retrospectively it can be difficult to prove other than by fitting what you know to be a suitable matched assembly for the given dish. Figure 3.9 illustrates an installation where the dish and feedhorn have been supplied as a kit, right down to a plastic housing for the LNB and depolariser.

Figure 3.9 *A matched dish and feedhorn*

Feedhorn failures

OK it's a bit obvious but we'll say it anyway. The top two problems with feedhorns are leaks and bugs. There should be a plastic cap on most modern, quality feedhorns. If this splits due to ageing or physical damage from flying debris, it will let water in. There should be a transparent membrane between the feedhorn and depolariser (or LNB if latter is not fitted) but this often gets left out by installers and thus water ingress can have significant results – damage to all three components can result. A severely corroded feedhorn where damage has resulted to its shape does not perform properly although results from good level signals will usually be unaffected.

If the feedhorn cap is damaged or missing, it isn't only rain that will find its way in! Varying and gradually reducing signals will often be found to be due to spiders or the like having nested in the feedhorn. As they move around so the signal varies! A largish spider is quite an attenuator at SHF. Remove all wildlife, replace the cap (or fit one if the original design was without) and smear some smelly lubricant around to put them off returning.

Multifeed systems

There are a couple of possibilities for systems to utilise a fixed dish to receive more than one satellite in more than one geostationary position. A fixed bracket that fits to the end of the dish boom and supports a pair of LNBs is a popular arrangement. The close positions of Astra and various

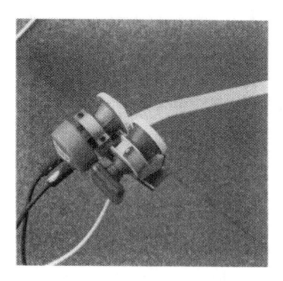

Figure 3.10   *A fixed twin LNB bracket viewed from rear*

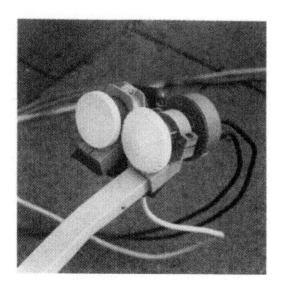

Figure 3.11   *A frontal view of the bracket in Fig. 3.10 showing the problem of LNB proximity when satellites are close together*

Eutelsat craft make this a relatively cheap and simple option (see Fig. 3.10). The 'extra' LNB uses signals from part of the dish and so the strength is not as high as that of a dish pointing straight at the satellite. A lot of playing around with position on the bracket and dish position (compromising azimuth between two desired satellites) will optimise results but the whole affair is a compromise. Do not waste hours trying to get minor improvements from such a system. The other problem here is that if your satellites are very close together, say 3°, there is not room for two LNBs side by side if they are both of conventional feedhorn size. Figure 3.11 illustrates how close together two such LNBs are for two satellites 6° apart. The answer to this problem is to use one polyrod lens LNB which has a very slim feedhorn. The cost then becomes a factor.

An alternative to using a fully fledged motorised system is to motorise the LNB. IRTE have produced a very elegant unit, the Multi sat which again bolts to the end of the dish boom. It utilises the existing LNB and coax cable and moves the former from left to right to receive adjacent satellites (to a maximum of 15° or so) in the manner of the fixed bracket above. It allows positions to be stored and very fine tuning of position. The advantages are clear – only one LNB and thus input to receiver, many satellites rather than two and in practical terms the results are far superior. The down side is cost, of course, compared to the bracket.

Cables

It will depend on your pre-satellite background as to how you view cable problems. Aerial installers will largely appreciate what goes wrong with coaxial cables but need to assimilate that with satellite television signals. Others, unfamiliar with terrestrial TV and radio reception will need to consider failures of the cables *and* how this affects satellite i.f. signals.

One needs to remember that the signals coming down the coax from the LNB to the receiver are already block down-converted. No tuning has taken place, but the whole band of transmitted and received frequencies has been down-converted as a group to a range of frequencies above the u.h.f. television band. This is known as 1st i.f., the exact range of which will depend on the LNB and satellite in question. In certain situations, satellite i.f. can overlap terrestrial u.h.f. frequencies (we shall see the significance of this later).

Another factor to bear in mind is that as well as conveying the 1st i.f. from outside to in, a d.c. voltage has to be passed from the receiver inside to the LNB for powering. It will in most cases be varied, typically switched between 13 and 17 V, to provide depolarisation control. It is vital therefore (as indeed with terrestrial masthead amplifier arrangements) that the down lead and its terminations are sound, good electrical and mechanical joints are made.

It is possible with u.h.f. signals to have poor contact in plugs (or indeed open circuit screens) and still get signal at the end of the lead. However, where a d.c. is required – satellite or terrestrial masthead – this will not happen. Where, as on some receivers, standard coaxial plugs are used, they must have the centres soldered and screens firmly clamped. Crimping with pliers or cutters is not acceptable. Apart from the possibilities of intermittent connections occurring, some poorly manufactured plugs have plating on the centre pin that peels off as soon as it is distorted by crimping. The result is a short circuit, possibly intermittent.

With just signal, the result would be a loss of signal. With a satellite receiver, the results could be much worse. If the unit is not well protected, as many older designs were not, power supply failures within the receiver will be the result. See Chapter 10 for more details.

Cable or interconnect failures can often be the result of a poor installation. Being aware of this can often save time in diagnosis.

Cable failures

What can we attribute to cables in the way of faults? The most obvious – and perversely least common – are those of open circuits and short circuits (see Fig. 3.5 a and b). Disconnecting the lead at the receiver end and placing an ohmmeter across it will normally show a high, often infinite resistance across centre and screen. This is of course reading not only across the cable but also the input to the LNB. A low reading (less than several kΩ) should arouse suspicion. It is often the case that suspects will be checked in a slightly illogical order. Here, for example, check the connector for a short and then disconnect the LNB and remeasure the cable. This will prove whether the short is in the cable or the LNB. The obvious reason for this is that the LNB is outside and usually requires the use of steps or ladders to reach it. Let common sense and human nature (or laziness!) come in to play here!

If you suspect an open circuit, you will need access to both ends of the cable anyway, and for a foolproof check on the cable, connect it to the receiver and measure for a d.c. present at the LNB end when disconnected from it – i.e. measure across the F plug. Alternatively, short one end of the coax (with the receiver obviously disconnected from it!) together screen to centre and measure for a short at the other end. A short reading indicates that the coax is not open. The current drawn by a typical LNB will not exceed a couple of hundred milliamps normally. This may increase where other items are powered via the feed.

Many receiver power supplies will start to pump (cyclically start and shut down quickly) when a short circuit is detected on the down lead and so simply unplugging the coax from a pumping receiver is a good primary check. If the receiver then comes on, look for a short with the coax or LNB.

The cause of down leads developing open or short circuits will often be due to them having been damaged. If clips break and they start moving in the wind, they will chafe on bricks or mortar or even become sliced through. Water will ingress via the smallest of openings and the braid and air-spaced dielectrics will promote rapid distribution of the water along the length of the coax due to capillary action. A tiny amount of water will go a long way in this situation. Water in the lead, however, will not often cause a short circuit. It will, combined with the exposure to air caused by the cuts in the outer sheath of the coax, cause corrosion, especially of the braid which is the first element it encounters, and assuming that it is a copper cable, the braid will turn black and begin to disintegrate leading to an open circuit. Water ingress will also lead to another, more common failing with coax down leads.

Mismatches

Far more likely a problem than a simple open or short circuit is a mismatch. This is where the impedance of the cable is not maintained throughout its whole length. Normally with most satellite systems this will be 75 Ω as for terrestrial TV and FM radio aerials. The result of a mismatch is that reflections occur in the cable at the point where damage has occurred. This damage is usually a kink or compression resulting in deformation of the cable (Fig. 3.12). The defor-

mation causes the impedance of the cable to change at this point and thus reflections occur within the cable – the degree of the mismatch and the specific measurements of the deformity will determine exactly what the effect is. What can be said is that the effect is likely – with analogue signals – to be very frequency specific.

Symptoms of a mismatch

The symptoms when you are aware of them are quite distinct but go unnoticed by many. The problems are usually that certain channels are affected. They are sparkly (both black and white) due to their level being low. Unlike overall weak signals, channels of higher and lower frequency will appear perfectly OK. Check the nearest channel(s) of opposite polarity to that worst affected. Are they similarly afflicted? If so, it looks like a mismatch. The symptoms can be so severe as to all but completely 'notch out' a couple of channels leaving all others around them perfectly OK.

The incidents caused by mismatches are quite great in number but some people fail to spot them. This is because, with mild cases, the symptoms are misinterpreted as a low gain or noisy LNB. Replacing it gives a sufficient change in characteristic to the system as to mask the mismatch symptom and apparently cure the fault. However, it won't have been cured.

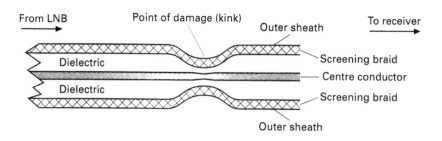

Figure 3.12   *Illustration of a cable mismatch due to physical damage*

How do mismatches occur?

We've mentioned that water ingress will cause a mismatch – and probably ultimately an open circuit screen – and that compression or kinking of the cable, as shown in Fig. 3.11 will. Let's look in more detail at the compression and kinking of cables.

Remember my belief that the majority of such problems are ultimately caused by poor installation? Here's a list of how mismatches occur. Use this as a checklist to make sure that you don't cause one during installation and as a means of finding the mismatch having decided that it is the likely cause of your problem. Also, see Fig. 3.13.

1. Kinking. This invariably occurs where the cable has to be bent. Spool the cable off its drum correctly. Do not pull off more cable than you need at a time, have the drum turning on an axle so that as you pull the cable (with minimum force) it does not become twisted.

Leaving the drum sitting on the floor to thrash around like a dying beast when you pull the cable is asking for a twist or kink to occur. As you run the cable, do not pull it tightly around any bend. This means around the dish brackets/arm and boom, and around corners in brick work or other building structures. When running cables, do not bend them tighter than the minimum bend radius specified by the manufacturer. A general rule is a minimum bend radius of ten times the cable diameter.

2. Crushing. A cable will be crushed when it is kinked but it can be directly crushed in a number of ways. Firstly, use the correct dimension of connector for the cable. Again, many do not appreciate that the almost universal connector, the F connector, is available in several different girths for different cables. It is also available as a screw-on or crimp-on device. Screw-on versions are very straightforward to fit and require no specialist tools. If the correct size is correctly fitted, they

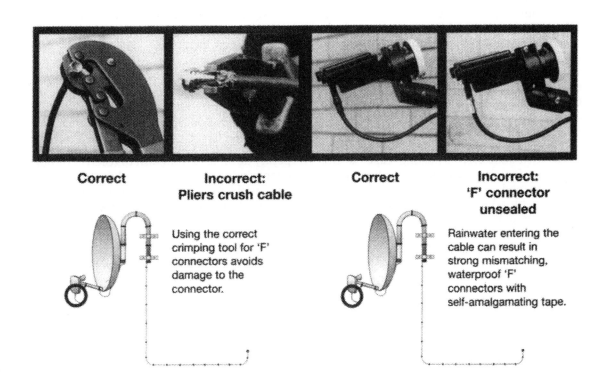

Correct   **Incorrect:**
Pliers crush cable

Using the correct crimping tool for 'F' connectors avoids damage to the connector.

Correct   **Incorrect:**
'F' connector unsealed

Rainwater entering the cable can result in strong mismatching, waterproof 'F' connectors with self-amalgamating tape.

Figure 3.13   *Astra's advice on avoiding mismatches (Astra)*

are excellent (although certain satellite companies do not advise them). If you use the wrong size, you will either crush the cable or provide a poor connection that falls off in time. Crimp-on connectors are fine as long as these constraints are borne in mind and that they are fitted with the correct crimping tool. It is absolutely unacceptable to fit them by squeezing with pliers!

Another way of inadvertently crushing the cable is to over-tighten cable ties. They should not make an impression on the cable. Similarly hitting the cable when clipping or over-doing the clips such that they flatten the cable. Always use the correct sized clips – too small and this crushing is inevitable, too large and the cable will move in the wind with the attendant problems of damage occurring. On a vertical run, this means that all the weight of the cable drop is being supported by the hanging point at the top thus causing kinking.

3. Water ingress. We've already discussed cable damage letting in water but some installers invite it in! The F connector at the LNB must always be sealed. Many arguments have ensued over the years about the best way to do this. The widely accepted best way is to use self-amalgamating tape (Fig. 3.14). There are now silicon based sealants that can be applied to the joint which do not liberate acetic acid during their curing process. Ones that do should be avoided as the acid enters the F connector and can damage the PCB inside the LNB. Self-adhesive, insulating tape is not adequate for the purpose of sealing an F connector and grease will leave you a fine mess to deal with when you have to work on the installation in the future.

Remember when routing cables to leave a drip loop at the LNB and where entry to a building occurs.

4. Splits and joints. It is amazing how non-technical people will just split coax leads to feed other rooms or join them to extend them. Even supposed professionals have been caught doing it (hence the term 'electrician's split'). The problem is serious enough at terrestrial level but when you consider the higher frequency and d.c. requirements of satellite 1st i.f. you can see that these practices have to be banned. There are commercially available distribution amplifiers and 'magic switches'

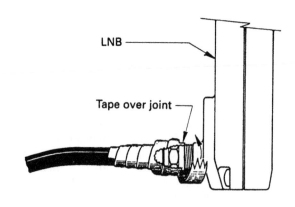

Figure 3.14 *Sealing an LNB F connector with self-amalgamating tape*

which take care of polarity control for Marconi systems and these should be used where necessary (see Chapter 5).

What to do

If you believe that you have a mismatch and you are not in possession of a spectrum analyser, then the most productive thing to do is to run a temporary down lead from dish to receiver through an open window. Connect up and see if you still have your problem. The spectrum analyser would be able to display the notched out area of the spectrum just to allow you to see on a trace what you can see on the TV screen. A similar test at the LNB output should show no such loss. This is all rather over the top for a single feed domestic dish, but when you have an i.f. distribution system (see Chapter 5) supplying many properties over a large area, a spectrum analyser is invaluable for this kind of fault tracing.

If you have cleared the fault, then change the down lead avoiding introducing any new faults! It is a complete nonsense to change a down lead with all the drilling and routing that involves without having confirmed absolutely that it is the cause of the problem. One vital point to remember is that you should *never* join a down lead. It doesn't matter how difficult it may be to replace part or how sure you are that the problem is only in one section, the whole down lead must be replaced. A joint is another mismatch waiting to happen!

Cables to use

There is in the industry a majority opinion that the ideal coaxial cable for i.f. down leads is CT100. This is actually one manufacturer's type number that has come to be the generic name for twin screened coax of standard 6/7 mm dimensions. As well as having a good heavy braided screen, it has a copper plate screen over the top of the braid. The overall package of build quality, performance and cost make it an obvious choice for standard i.f. down leads. Therefore, if you need to replace a down lead, using CT100 or one of its equivalents will see you right. Using a good quality, well braided single screened coax will also rarely cause problems. What must be avoided is cheap, poorly screened, lossy coax cables. Check the manufacturer's specification. The attenuation per 100 m of CT100 is 21 dB at 1000 MHz rising to 28.3 dB at 1750 MHz. Compare this with what you are considering using. Genuine CT100 has been evolved to handle DTH satellite i.f. signals and meets the current standard BS EN50117. In terms of screening, CT100 well exceeds the required 85 dB screening attenuation of Cenelec TC209.

Where more than a coax is required, there are a number of purpose-designed cables available. For a fixed system which does not use a Marconi type LNB, but a servo motor or electromagnetic depolariser, then a single encapsulation coax with three or two core cable is used. Where a motorised system is concerned, two extra wires are required for the actuator motor drive and two more for feedback (Fig. 3.15). These cables tend to be directly buriable. If, however, you get water in them, they can be quite hefty to dig up and replace! If the water damage was purely in the coax which is sealed from the rest of the conductors and the cable was in a trunking, it may be possible to feed a new single coax to replace it rather than replace the whole larger cable.

Terminations

With modern LNBs and receivers, the standard i.f. connector is the F connector. There are, however, some exceptions to this. From an LNB point of view, the only real alternative is the N connector and this is very unusual. Receivers have used a few different connectors. The standard male or female coax connector is sometimes used

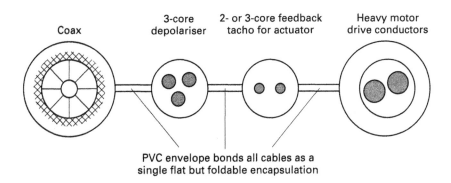

Coax   3-core depolariser   2- or 3-core feedback tacho for actuator   Heavy motor drive conductors

PVC envelope bonds all cables as a single flat but foldable encapsulation

Figure 3.15   *Layout of cable for a motorised system*

– sometimes both on a single receiver with two i.f. inputs. Alternatively a BNC connector may be used on older units. Whatever is used, it is important, as we have said, to ensure mechanically and electrically sound connections and so fitting of plugs correctly is essential.

The F plug uses the centre conductor of the coax as its centre 'pin' – i.e. there is no connection within the plug made to the centre coax conductor – it simply pokes through the body of the plug. It is important to trim the centre a few millimetres proud of the body of the F plug to allow for locating. Similarly it should not be too long otherwise the plug will not screw on fully without bending the centre, causing a short circuit in the plug, or by forcing the centre through the back of the socket, causing a short in either the LNB or receiver.

Astra 1D interference

This problem is one which came upon the industry rather rapidly and now is likely to be encountered rarely – although a similar scenario has the potential to occur again in the future with down conversions of other satellites' frequencies. The problem was simply that a system, working perfectly normally with a receiver and dish using an LNB with a local oscillator of 10 GHz (standard for all Astra 1A–C installations), down-converted Astra 1D frequencies when they appeared – although the vast majority of the receiver/LNB combinations couldn't tune to these frequencies. This meant that their first i.f. sat over the top of the u.h.f. terrestrial TV band (Fig. 3.16). Therefore in areas where the terrestrial TV signals were at the upper end of the band (group C/D) any crosscoupling of this low first i.f. would cause interference, in the form of patterning or cross-modulation, to respective terrestrial channels. This gave the bizarre scenario of customers with no satellite system complaining that their BBC1 was being wiped out and they could see an Astra 1D test card in its place! The cause of this problem was the satellite down lead. Bear in mind all that has been said about the quality and performance requirements. Some of the coax used in early installations was very poorly screened and literally leaked signal. Combine this with the very real possibility of equally poor down leads on u.h.f. aerial installations and the potential for problems is very high. One poor installation in a street could affect many other properties' signals.

Down leads should have been changed but in many cases weren't. A filter was available to simply insert in line to block all the lower, interfering frequencies and in other cases the LNB was changed to a '1D' type – i.e. one with a local oscillator of 9.75 GHz which moved the first i.f. back up. This caused other problems though with tuning. We discuss this in Chapter 4.

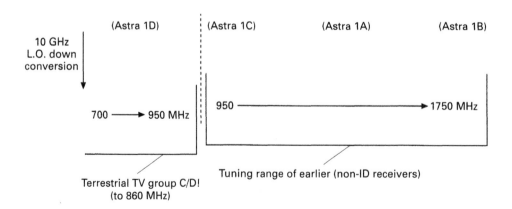

Figure 3.16   *Interference from Astra 1D when uing a 10 GHz LNB and poorly screened cable*

Down leads and digital signals

As we begin to see an increase in the equipment that processes digital signals, it will begin to dawn on people that digital systems give much different results to analogue ones for a given failure. We have seen it with digital video systems but as yet many have not experienced digital broadcast reception problems. Digital systems tend to either work or not. They do not gradually degrade. They will tolerate a lot – they are robust with remark- able error correction abilities – and continue to work, apparently perfectly up to a point. Beyond this they will not operate and simply stop working. Thus the analogue system clues of sparklies or audio noise are probably not going to be available to you for fault finding. Specialist test equipment, allowing you to view the data-stream and measure its error rate will be the order of the day.

One very important thing to understand though is that down lead problems, especially mis- matches, will result in no signals – beware!

4

DEPOLARISERS AND LOW NOISE BLOCKS

These devices, whether a single unit or separate, form what is known as the head end of a dish installation. These are the electronics sitting at the end of the boom (although not necessarily in front of the dish). When considering servicing and reliability, one has to consider the overriding factor that these units are normally operating in rather adverse conditions for electronics – in the cold in winter, and heat during the summer, the temperature range can be great. Add to this the possibility of extreme rain and wind and you cannot really blame them for the odd failure! In our list of likelihoods of failure then, it is no surprise that we count these as higher than receivers when it comes to signal faults.

Depolarisers

Or as the world wishes to call them, 'polarisers', which is incorrect as you will see if you don't already appreciate it. As with terrestrial signals, to obtain maximum use of the ever-valuable bandwidth in which they have to broadcast signals, satellite service providers transmit half their channels in the opposite polarity or plane to the others, thus allowing them to be much closer together than would otherwise be possible (Fig. 4.1). While terrestrial signals are always of the same polarity from an individual transmitter, this is clearly not the case with satellite – at least not where orthogonal (horizontal or vertical) polarisation is used. In the vast majority of cases with DTH satellite, orthogonal polarisation is used and that is what we shall be considering. The possibility exists to use circular polarisation where the signal is polarised either right hand or left hand – clockwise or counter-clockwise. In these circumstances (usually with nations' DBS satellites), the direction tends to be the same for all services on that satellite but the opposite is used on adjacent satellites. This means that a fixed depolarisation

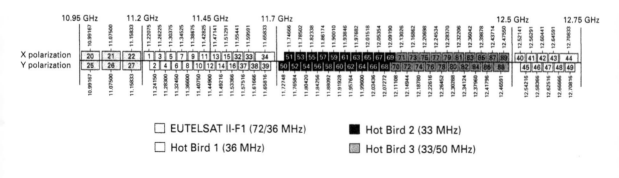

Figure 4.1 *Transponder map showing the use of two polarities (Eutelsat)*

system can be used in the receiving system – usually a slab mounted at 45° to H and V in the throat of the feedhorn.

To be able to obtain all channels on both horizontal and vertical polarities, we thus need some form of depolarisation. If we mount our LNB such that its dipole is oriented for vertical transmissions, then to receive horizontal ones, we need only twist them through 90° before presenting them to the LNB. By doing so we will also twist the vertical ones by 90° making them horizontal and thus preventing interference from them. This is why the depolariser appears between the feedhorn and LNB.

Types of depolariser

There are three types of depolariser for orthogonal systems.

Electromagnetic depolarisers

Here we have a device separate from a conventional LNB. A current is passed through a coil wound onto a ferrite core. The waveguide between feedhorn and LNB passes through the

centre of this and so as current flows an electromagnetic field is produced which twists the signal by an amount proportional to the current flowing. This then provides a very controllable depolarisation system and, unlike the Marconi type, by virtue of this can be used on multi-satellite systems. Consider the fact that as a dish moves in an arc to position itself for satellites in different geostationary positions, the definition of horizontal and vertical with relation to the earth varies. Thus at each dish position, if not each channel, we need to be able to vary the twist applied to the signal to match it to the LNB dipole (Fig. 4.2). By varying the current through the coil of this type of depolariser we can achieve this twist – it is called skew adjustment and can usually be set and stored for each programme position.

The electromagnetic depolariser requires two conductors extra to the coax to provide the current for operation. Thus a twin cable is used. There are a number of designs which incorporate a feedhorn with the electromagnetic depolariser (Fig. 4.3).

Figure 4.2

Figure 4.3 *An electromagnetic depolariser combined with feedhorn. The twin leadouts can be clearly seen*

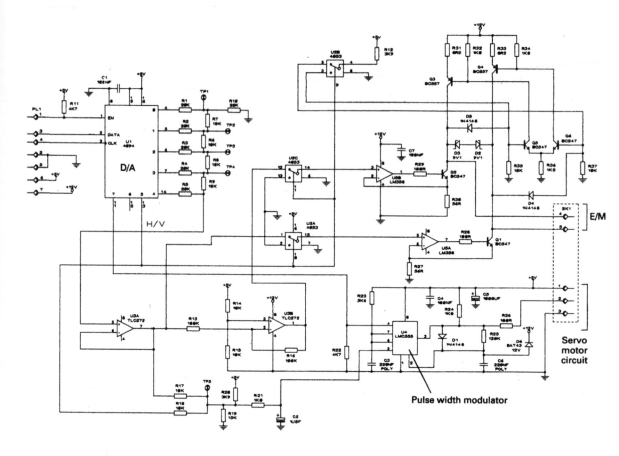

Figure 4.4   *Drive circuit for an electromagnetic depolariser*

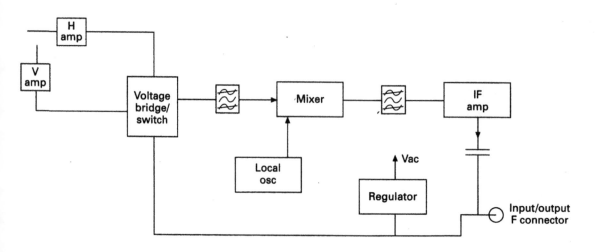

Figure 4.5   *Block diagram of 'Marconi' type LNB*

The drive circuit within the receiver usually provides 12 V or 5 V with a swing of either ±50 mA or 100 mA. Figure 4.4 illustrates an example circuit. Problems here tend to be due to loss of supply or short or open circuits in the transistors causing, one way or another, a fixed polarisation skew. Remember that failures here could cause weak or no signals on either polarisation dependant on how much current is flowing and what the null orientation is. When the depolariser is used with a motorised system, it is generally configured such that with no current flowing it has an equal swing to H and V with the dish pointing south, thus current needs to flow to achieve either polarisation. Therefore if the drive failed, there would be at best weak signals on both polarisations. As the dish moves to either end of the arc, one polarity will be achieved with no current flow and the other will need more, and vice versa, at the other extremity. The accuracy of these statements will depend on the starting orientation and the resistance of the depolariser coils. With a fixed dish system, no depolarising will invariably lead to loss of one polarity only.

This d.c. resistance is typically from 60 Ω to 100 Ω – sometimes beyond either of these limits but nominally 80 Ω. Therefore a simple resistance test can be carried out on the depolariser and its cable from inside. With certain designs the connections to the cable were extremely prone to corroding away thus causing open circuits and no depolarising. Like F connectors, these joints should be sound and completely waterproofed.

When working on a receiver with a problem in this area, when aligning it – or indeed just checking the circuit's operation – it is not always practical to have a dish fitted with such a depolariser. It is therefore practical to fit a suitable, say 80 Ω resistor, across the two terminals and measure across or in series with it. There are, with some circuits, current adjustments to calibrate for e/m drives.

Voltage switching (Marconi)

By far the most common system today is the voltage switching type or what is commonly referred to as 'Marconi' type. The depolariser and indeed feedhorn are combined into a single unit with the LNB with this design and polarity is switched by simply changing the d.c. voltage from the receiver to the unit from 13 V for vertical to 17 V for horizontal. Figure 4.5 gives a block diagram of this design showing that a simple bridge network is used to switch between the two front end stages. The obvious advantage of this system is simplicity in both cabling – only the coax is required – and receiver design.

In the receiver, system control simply switches the LNB supply output via the tuner between two supplies. This can be seen in Fig. 4.6 where logic, via Q106, switches different resistances in the control leg of programmable regulator IC100. It regulates either 17 V or 13 V from the 20 V supply at its input and this is injected via the tuner. An alternative design from an older receiver can be seen at the bottom of Fig. 10.5 later in the book where Q31 and Q3 are used.

Problems that can occur are usually based around loss of one polarity. The first thing that has to be done is to establish whether the problem is with the depolariser within the LNB or with the receiver. A d.c. voltage check on the dish connector of the receiver should answer this. Toggle between horizontal and vertical and confirm that the voltage switches between 17 and 13 V. If not, then you have a problem with the receiver. The likely cause of such problems is in the switching circuits – i.e. Q100 or the regulator IC100 in Fig. 4.6 or Q3 and less likely Q31 in Fig. 10.5. The logic level from system control is rarely in trouble.

If the receiver is OK then a replacement LNB is the order of the day. Because of the way in which it works, you cannot simply connect more than one receiver to a single LNB and get it switching polarities correctly. There are diode splitters available but if either receiver has 17 V selected, then this will clearly override 13 V from another receiver as they will be in parallel. It is therefore important to make sure that you test a system with a switching fault by connecting the receiver directly to the LNB, thus avoiding any misleading problems caused by DIY distribution systems. Proper ways of attaining multi-point satellite are discussed in the next chapter.

LNB polarisation offset

Because of the factor mentioned earlier about the orientation of horizontal and vertical varying for different azimuths, when a Marconi LNB is installed, it should be twisted off centre to achieve

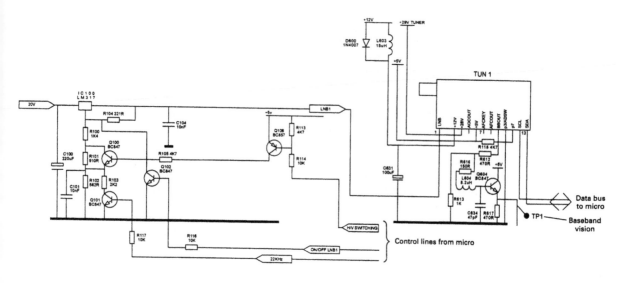

Figure 4.6   *Circuit showing H/V switching for a Marconi type LNB (Grove Farm Publications)*

matching with this 'error'. If this doesn't happen, correct switching and allocation of the two polarities will not occur. These problems can be very intermittent and range from poor results on one polarity to no signals at all on channel change (see Fig. 4.7). In the UK, to receive Astra, a clockwise offset of around 7° is necessary, as viewed from front of dish. Many LNBs have values engraved in the top of the shoulder to enable this to be set easily. Some LNBs have offset built in – check the specification so as not to be caught out.

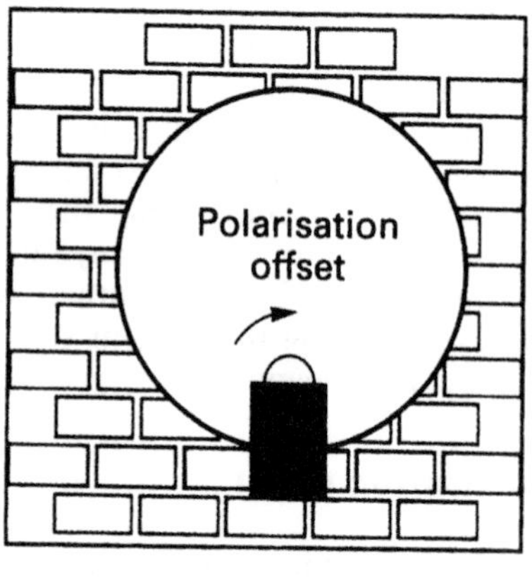

Servo motor depolarisers

These are also known as polarotors, especially in America. They are now virtually obsolete in the UK due primarily to their cost, reliability and insertion loss (over 3 dB compared with 0.5 dB for electromagnetic). They work on the basis of a motor turning a resonant probe in the throat of a waveguide between the feedhorn and LNB. One can see the immediate potential for problems with the motor, and indeed one wouldn't usually be disappointed! The driving of such a device is via a three core feed: a d.c. supply, usually V, ground and a pulse width drive which switches the supply across the motor. One sets the pulse width at the receiver to determine how far the probe would be turned – one is set for H and one for V and obviously there should end up being 90° between them. The pulses (and sometimes supply – which can cause problems if using an interface for say an electromagnetic device, see below) are applied for a fixed period of time and thus the width of the

Figure 4.7   *LNB polarisation offset (Astra)*

pulses determines how far it moves, i.e. how long the motor is on. Due to their mechanical nature, their operation is unreliable when old – not returning to exactly the right place, etc. – and the best idea is to replace them with an electromagnetic device.

Depolariser interfaces

It is not always possible to directly replace one type of depolariser with another, or maybe you need to install a receiver that cannot drive, say an electromagnetic unit that is already there and working perfectly. There are available a series of interfaces to convert from one type to another. The most popular of these is made by Global Communications – see Appendix 3. Some receivers are available with retro-fit options for alternative depolariser drives and some come complete with the ability to drive some or all.

The low noise block (LNB)

There is another satellite television term that has become rather too wide in scope. The LNB consists of a low noise amplifier (LNA) and a low noise converter (LNC), the latter consisting of a fixed local oscillator and mixer. Its purpose is to down-convert the off-dish frequencies (10–12 GHz for a typical Ku band system) to a first i.f. of around 900–2000 MHz. This is a standard superheterodyne principle with the subtle difference to a tuning system, that all frequencies in are down-converted as a block, no tuning takes place. The point being that such high frequencies as those off-dish simply would not appear at the other end of a cable were you to try to send them down it! Similarly it is far easier to amplify and process a range of lower frequencies from the receiver's perspective. The signal in is via a waveguide (electromagnetic) and the down-converted output (electrical) via a connector (usually F type) to a coaxial cable. The d.c. power supply for the LNB is fed up from the receiver via the same coaxial cable. This then is the extent of the traditional LNB (see Fig. 4.8). This is the form of LNB that will be present on a motorised system and/or interfaced with either a servo motor or electromagnetic depolariser and a feedhorn. The term LNB is used in modern parlance to describe the

Figure 4.8   *A traditional LNB*

all-in-one 'Marconi' unit which is in fact an LNB, depolariser and feedhorn assembly in one.

The design of an LNB is rather different to that of a conventional electronic circuit as both electrical and electromagnetic signals are present. The oscillator is a resonant cavity type meaning that the internal dimensions of the device and the specific layout of the circuit and the physical casing are all vital factors in its operation. One can draw analogies with magnetrons and Klystrons. Figure 4.9 gives an example of an equivalent electrical circuit for a conventional LNB.

Local oscillators

This book predominantly concerns itself with Ku band satellite equipment – in many cases, there is little difference between Ku band and other band equipment. Clearly where LNBs are concerned, the frequencies involved and therefore dimensions are going to be rather different. Here we again discuss specifically Ku band devices. The typical local oscillator frequency of such an LNB will be 10 GHz. Most mainstream receivers prior to the launch of Astra 1D assumed that they would be connected to such an LNB. All mainstream receivers intending to drive Marconi LNBs produced in the years since 1994 will have the facility to at least be able to work with LNBs with 9.75 GHz and 10 GHz local oscillators.

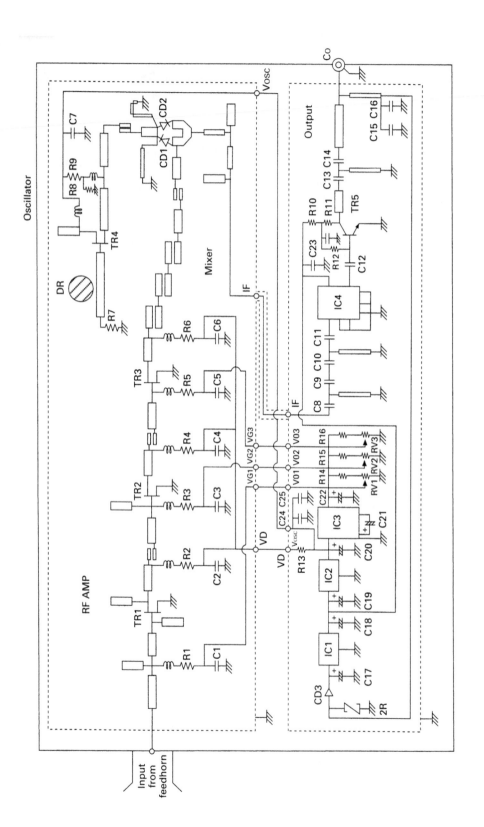

Figure 4.9   An electrical equivalence circuit diagram of a traditional LNB

I.f. offset

More advanced or specialist receivers had the facility to dial in the specific oscillator frequency to ensure compatability. Where oscillator frequencies are preset (be it switchable presets or not) as in Marconi type receivers, there will be a 'fine tune' adjustment to match the tuning scale of the receiver with the local oscillator of the LNB. This is known as the i.f. offset and is obviously an overall adjustment which affects all channels – the effect of misalignment being that the tuning points are all shifted. For example, there will be black sparklies or white sparklies on all or most channels when they are set to the correct frequency due to them being slightly above or below the tuning point. On many receivers this adjustment has to be set manually and stored. The best approach is to use a weaker channel or one with a narrower offset such as UK Gold or MTV on Astra or BBC World on Eutelsat 2F1. If optimised on these channels, it will ensure greatest accuracy. On certain receivers, there is an AFC facility on the i.f. offset. This generally saves problems but if the LNB offset has not been set correctly or the local oscillator in the LNB is well out, then some more drastic symptoms can occur such as signals disappearing on channel change. The AFC can be disabled to allow manual adjustment and operation.

The Astra 1D scenario

This fourth satellite was launched and as a result of the success of the three previous craft, and thus demand for transponders, used for DTH transmissions in an area of the band that was not previously allocated. This meant that virtually all receivers marketed to that date were not able (as they stood) to receive signals on those frequencies (see Figs 4.10 and 3.16). An LNB with a local oscillator at a lower frequency (9.75 GHz) was required with a tuner and tuning system in the receiver with a wider bandwidth (950–2050 MHz first i.f.). There were, however alternatives to this (effective) requirement for a new system.

A device named the ADX plus by Global Communications can be used to translate (up-convert) the Astra 1D band first i.f. into the area of Astra 1A first i.f. and thus into the range of any standard receiver. It can be used in this mode with a 10 GHz l.o. LNB and will give varying degrees of coverage (usually not down to the lowest frequencies) of the 1D band. By switching the converter internally, one can use it with a 9.75 GHz LNB where it also serves to down-convert the B band which will then have gone 'over the top' of the tuning range of the standard receiver. The converter is thus switched in when required in each situation. This can be remotely controlled by use of the TV/SAT button on the remote handset and switching status line on pin 8 of the TV scart of many receivers (Fig. 4.11). All in all, this is a superb device.

The inclusion of this device or similar units can give rise to some confusion. By virtue of its up- and down-conversion abilities with a 9.75 GHz LNB, there will be two tuning points for the same channel (for certain channels/frequencies) one when the unit is switched in and one when switched out. Customers can get confused in that if the unit is switched in and they change to a 1A channel, for instance, it will not be present or may be a different 1D channel – this is simply education. The overriding factor to bear in mind,

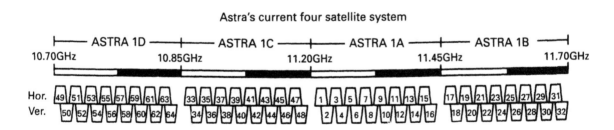

Figure 4.10   *The Astra 1D band in perspective (Astra)*

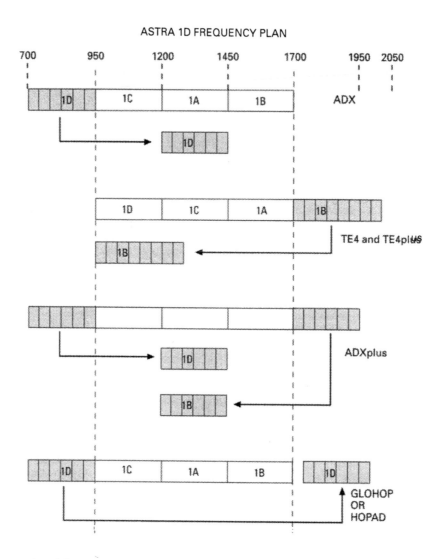

Figure 4.11   *The operation of the ADX plus and equivalents (Global Communications)*

however, is that frequency readouts on receivers become meaningless – i.e. they will be either 500 MHz or 250 MHz out when the unit is switched in.

LNB faults

When you consider the extremes of temperature that the average LNB is exposed to, its complexity and the fact that the oscillator must remain stable

for the whole affair to work, it is fair to say that they are very reliable. Like very large integrated circuits, many LNBs get changed when they are not at fault – I hope that as a result of reading this book, engineers will realise this and help reduce the incidents to a minimum. The traditional style LNB is possibly one of the most reliable electronic devices made – its relatively low-scale production and higher cost make them more reliable it seems – that is allowing for the significantly lower number encountered.

The Marconi type devices are also very reliable on the whole. There have been certain designs by certain manufacturers which have been intolerably unreliable – with *genuine* faults. Again the pressures of large-scale production seem to have taken their toll.

What fails?

Where LNBs genuinely fail one can usually connect the failure to the weather. Water ingress is an obvious problem, via the F connector or where a cable has been cut and water made its way into the LNB are the two favourites. In such cases, replace the LNB and cable. Ensure that the new ensemble is correctly waterproofed.

A complete loss of signal output due to faulty LNB (refer to previous chapter and fault tree) is often preceded by very cold weather. In the early days, it was felt that an LNB may suffer from the sun concentrated upon it by the dish but in fact this doesn't seem to have been a problem. A dead LNB can be confirmed by measuring the current being drawn. This is a facility possible on most good self-powered signal strength meters. Most modern LNB designs will normally draw around 200 mA although older ones (including Marconi blue cap) may run up to 300 mA. If, however, you get a very low current drawn, then you can safely assume the unit to be faulty. Similarly an excessively high current drawn will indicate a leak or short in the unit. Remember that too great a current drawn by the LNB will usually cause an indication or failure with the receiver.

Low output causing weak signals is the most often encountered problem. One needs to have an idea of the expected level. In a typical 80 cm dish, non-universal LNB, Astra installation a signal level out of the LNB between 80 and 90 dBuV is to be expected. Assuming that you have eliminated all other possible causes – dish off beam, receiver faulty, etc. – you can assume that the LNB has failed in this way and the definitive answer would be a replacement. However, if your problem is frequency specific or the fault varies, then it is unlikely that the LNB is the true cause of the fault. Do not be misled with in-line signal strength meters – when in circuit, they can give very odd symptoms akin to a mismatch.

Local oscillator drift can be a problem. Most receiver designs maintain the LNB supply even when they are in stand-by. This serves, by keeping the LNB powered, to maintain temperature stability therein. In designs which do not, or where a customer insists on unplugging the receiver when not in use, it is perfectly feasible that you will encounter tuning drift as the oscillator in the LNB 'warms up' each time the receiver is powered on. Similarly if drift occurs outside of these situations or if the i.f. offset needs to be set near an extremity, the oscillator in the LNB is suspect. If the fault is intermittent, soak test the system by installing a loan receiver to prove if you don't want to change the LNB on speculation (due to difficult access perhaps). There is, on certain designs, accessed via a bung on the body of the LNB, the ability to adjust the local oscillator, but to ensure accuracy requires very good test equipment.

If an LNB has been dropped either upon installation or otherwise, never trust it. The cavity construction can be very easily distorted or broken by a fall. Dispose of the device straight away.

Viability of LNB repair

The only real reason that anyone would consider LNB repair is out of interest (although repairs to low frequency LNBs are no more involved than repairing u.h.f. tuners, etc.). The cost of an LNB – especially the Marconi type – and the ability of the average engineer (without the required test equipment) to achieve an accurate result mean that as a commercial practice for service engineers it is not viable. It is a fact that many failures are simple – dry joints on F connectors, regulator failures, etc., but you cannot simply dismantle an LNB and put it back together again – remember that the whole thing works on the basis of tuned cavities (Fig. 4.12). Additionally, the cases are hermetically sealed and gaskets can be r.f. conductive. Therefore the order in which screws are put back in and the amount they are tightened will all affect alignment – especially that of the local oscillator. As we said above, if the local oscillator is off, so will be the tuning calibration. To recalibrate the oscillator requires a spectrum analyser, and to measure and test the gain and noise figure (say if you were looking for a low or noisy output) requires specialist equipment of considerable cost. Figure 4.13 shows the test bay at MCES in Manchester, England, which specialise in LNB remanufacture – see Appendix 3 for details.

Figure 4.12 *Two views of inside a 'traditional' LNB. The d.c. voltage regulator and local oscillator adjuster can be clearly seen.*

Table 4.1 *Specification table highlighting the criteria requiring consideration when selecting a (replacement) LNB.*

ELECTRICAL	UCJ 6520	UCJ 6534 Astra Single Output	UCJ 6549 Astra Twin Output Switched V/H	UCJ 6526 Astra Dual Output Fixed V and H	UCJ 6530 French Telecom Single Output	Astra Ultra Low Noise Block Single Output
Input Frequency range	10.95 – 11.70GHz		●	●	12.5 – 12.75GHz	●
Total Noise Figure	1.1dB typial (1.5dB Max)		●	●	1.2dB typical (1.5dB Max)	0.85dB typical (1.ødB Max)
Cross polar discrimination	25dB typical		●	●	●	●
Conversion Gain	55dB typical		●	●	●	●
Output Frequency Range	950 – 1700MHz		●	●	1025 – 1275MHz	●
Local Oscillator Frequency	10.0GHz		●	●	11.475GHz	●
Local Oscillator variation with temperature	≤ ± 3MHz		●	●	●	●
Gain ripple over any 27MHz band	± 0.5dB		●	●	●	●
Output impedance	75 Ohm		●	●	●	●
DC Supply	11.5v – 14.0v Vertical / 16.0v – 19.0v Horizontal		●	11.5v – 19.0v	●	●
Supply current	140mA typical (200mA max)		280mA typical (310mA max)	●	●	I
MECHANICAL						
Size	120mm x 66mm x 105mm x 25mm		●	●	●	●
Connector	1x F type Female		2x F Type Female	2x F Type Female	●	●
Fixing	23mm or 40mm ø		●	●	●	●
Finish	Zinc diecast, passivated and wet paint coat		●	●	●	●
ENVIRONMENTAL						
Operating	–30°C to +50° C		●	●	●	●
Storage	–40° C to +60° C		●	●	●	●

● = *Specification as per UCJ 6520*

Replacement LNBs

Clearly when we have decided on the need for a replacement LNB, we can simply fit a like replacement (new or remanufactured). It is the case, however, in the fast moving satellite market that the exact specification may no longer be available. The specifications will be better and so one needs to be able to interpret these relevant specifications and be able to decide the best type for the job.

When in the position of requiring a replacement, one should also consider the possibilities and benefits of upgrading the device. The cost of a universal LNB over that of a standard Marconi type with single local oscillator is minimal. Consider the likelihood of the receiver being upgraded in the near future and whether this justifies fitting one in preference. Some receivers capable of working with a universal will be found to have been installed with a 9.75 GHz device. Replacement with a universal may therefore be prudent. Conversely, with Astra systems, it is worth considering that digital services aimed at the UK will not be coming from its current orbital slot at 19.2° east but from the new slot at 23.5° east and the Astra 2 series craft. Dual feed systems are likely to proliferate but will you go to a universal LNB then or now?

The specifications to consider are as in Table 4.1. The obvious ones to get right are that the LNB is covering the correct frequency range and that the local oscillator is compatible with the

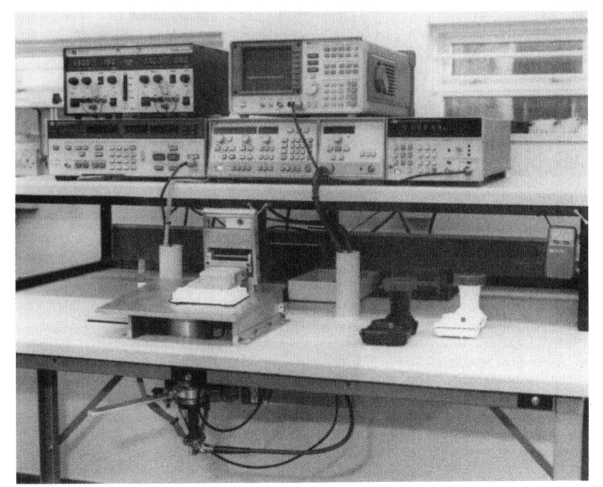

Figure 4.13   *The test and repair bay used for professional LNB remanufacturing showing test signal injection via the feedhorn and gain and noise figure measurement equipment (MCES)*

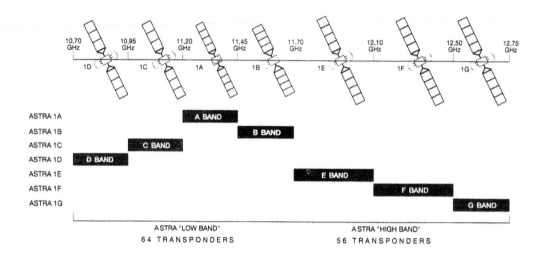

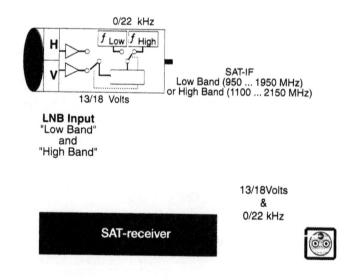

Figure 4.14   *The universal LNB showing block diagram of operation and construction and the high band which it serves (Astra)*

receiver – fitting a 9.75 GHz with a receiver capable of working only with a 10 GHz unit will cause significant tuning shift and loss of higher frequency channels. The other point to consider is the girth of the LNB where it must fit in the mount on the dish boom. There are collars available to help where LNBs are too small but if they are too big and you have no replacement saddle clamp, then you will not make it fit! Do not fit replacement clamps or mounts not designed for the specific dish – the angle and focal length are obviously critical to ensure that the LNB is looking at the dish correctly.

Noise figure

This is often over-hyped when considering LNBs. Clearly the lower the noise figure, the better, but we must draw the line where cost is no longer justified by improvement. If we are talking about fixed dish systems a noise figure in an LNB of around 1.0 dB should be more than adequate. If the results are still poor, the answer is to increase the size of the dish. This is an installation consideration. With traditional LNBs on a motorised dish working with signals of tremendously variable levels, then noise figures down to 0.6 dB are practical but as the noise figure goes down, the price goes up!

When looking at this from a service/replacement perspective, we can see that by replacing an old LNB – say a blue cap Marconi with a noise figure of around 1.8 dB, fitting a current Marconi type of perfectly standard performance and 1.0 dB noise figure – we are likely to improve the results overall, especially under heavy cloud cover conditions. Do not, however, expect to be able to go around solving poor signal problems by fitting lower noise figure LNBs – it won't work! If signals are low and there is no actual fault, the LNB does not possess a very high noise figure and there is no line of sight impedance, then the answer is a bigger dish!

Multi-band LNBs

There are conventional style LNBs available that have more than one local oscillator. This allows their use in more than one area of a band, or more than one band. Figure 4.1 illustrated the wide frequency range available at 13° east from Eutelsat. Assuming a suitably capable receiver (tuning range and i.f. bandwidth), the inclusion of a multi-band LNB would be necessary to cover all of these channels. (We need also consider the fact that some may be using MAC as opposed to PAL, or SECAM and some may be digital.) These LNBs may be two or three band and switch bands by varying the d.c. supply voltage – in much the same way as a 'Marconi' LNB switches polarity. The main causes of problems – aside from pilot error – are loss of one of the oscillators, necessitating a new LNB and loss of the voltage switching capability in the receiver.

The universal LNB

This unit, working on the Marconi H/V switching principle, is the latest breed of LNB. It is designed to enable reception of existing analogue servicing using a local oscillator of 9.75 GHz but also to be able to switch to a higher local oscillator of 10.60 GHz for the reception of signals in the so-called high band between 11.7 GHz and 12.75 GHz (Fig. 4.14). This is therefore a dual band LNB but the extra capability is designed specifically with digital transmissions in mind. The band switching is carried out by means of a 22 kHz tone which is sent from the receiver to switch the LNB from one band to the other.

The specific requirement of any satellite reception system intending to process digital signals is for minimum phase distortion. It is this property which is paramount therefore in any universal LNB. This is because – as can be seen from Chapter 14 – digital transmission systems rely on phase modulation techniques. Phase distortion would thus cause significant problems. It is very likely that older designs of LNB will not have sufficiently good phase characteristics for reception of the new digital services. Aside from the mismatches with cables already considered, faulty or just poor specification LNBs will be the main cause of problems with digital services in the early days.

22 kHz tone switching

Each channel preset on a suitably equipped receiver will have the option to set 22 kHz switching on or off dependant on which band the service is in – i.e. whether the higher local oscillator is required. This option will only appear if the receiver has been told that a universal LNB has been connected. 22 kHz tone insertion can be seen in the earlier Fig. 4.6.

There are available external 22 kHz tone generators and switches for retrofitting to older receivers – they may be used for a variety of purposes such as remote multiple LNB switching. These devices are again available from Global Communications.

DiSEqC (Digital Satellite Equipment Control)

A relatively new development in the switching field is rapidly finding its way into satellite receivers and ancilliary equipment to provide more powerful and flexible switching capabilities than plain 22 kHz. The DiSEqC system, positively encouraged by Eutelsat is effectively a communication bus as discussed in Chapter 11. This bus is conveyed via the single coaxial cable between dish and receiver. The good news is that the format is widely agreed and standardised across manufacturers. Being an extension and development of the 22 kHz principle it allows backwards compatability but its flexibility means that the signalling can be used for control of motorised systems or multi-satellite/LNB switching.

The bus master is the receiver's main microprocessor and ancilliary devices are slaves on the bus. Naturally the theoretical potential for problems is immense due to the added complexity of this control arrangement. In practice however the problems will relate to interconnections and basic set-up problems. There are available DiSEqC test jigs which can be plugged in line to decode the data protocol to aid fault finding.

5

DISTRIBUTION SYSTEMS

As with terrestrial television, it becomes desirable to have signals from satellite in more than one room of a house. When one also considers the rather more conspicuous antennae, one appreciates the need to do this by splitting the signal one already has rather than planting dishes all over the place. Equally one might, on a larger scale, wish to distribute signals to a group of flats or apartments within a large building. The generic term for these systems is SMATV (Satellite Master Antennae Television Systems). Beyond this scale we are into cable TV systems which are beyond the scope of this book but which include many of the principles discussed in this chapter.

Signal distribution methods

The most basic form of distributing a satellite TV signal is to simply split the r.f. output of the receiver and feed a cable to another room. This gives the obvious advantageous possibility of viewing the system in, say, the bedroom by just leaving the receiver tuned to the desired channel in the lounge. Naturally only the same single channel can be viewed at any given time. If the output is taken from the r.f. out of the VCR, it too can be viewed on the remote point.

If, however, we want to view more than one channel at any given time, we must use more than one receiver and so split the signal into them.

As with each subject in this book, detail of the theory behind this can be assimilated elsewhere. We do need to consider the basics, so that we can identify what type of system we have in order to determine fault finding approaches. From a signal point of view, we have to maintain the level, despite splitting it, and also, and very importantly, impedance. We mentioned in Chapter 3 the often encountered misdemeanours of the incompetent where cables are simply twisted together to form an 'electrician's split'. Where satellite is concerned, we have also to

consider the inclusion of d.c. supply voltages and depolarisation control.

There are two ways in which we can distribute satellite signals. The first, which is what would be used in a domestic situation, is i.f. distribution. Here we take the output of the LNB and distribute it at this frequency – i.e. typically 950–1750 MHz. The hardware may also allow simultaneous distribution of u.h.f. and v.h.f. signals to maintain a degree of simplicity and serviceability. Figure 5.1 illustrates such a system.

Otherwise we may translate the frequencies into the u.h.f. or v.h.f. TV bands and then use an up-converter at each point to translate back into i.f. for the receiver. This allows integration with an existing u.h.f. TV distribution system but means that bandwidth is limited. There is finally the possibility of simply receiving and tuning selected channels and then remodulating them onto an r.f. carrier in the TV band so that they may be tuned in like any terrestrial channel on a standard TV without any receiver in each property. Hotel installations are a prime example of this. The number of channels that can be supplied in this way is limited and the choice is seldom going to please everyone! Chapter 12 discusses choosing and implementing suitable u.h.f. output channels. Figure 5.2 illustrates this type of system.

The way that cables are run and thus the shape of the system from an interconnection point of view will vary depending on when it was installed and for what purpose. MATV systems (for terrestrial signals) that have since had satellite added will probably be different from those designed with satellite in mind. Similarly i.f. distribution to more than one property will have two cables to each extra to the terrestrial signal. Figure 5.3 outlines the possible layouts. Older systems may be loop wired where the coax loops into one point and then out to the next and thus any open circuit kills the rest of the leg. This really is not suitable today.

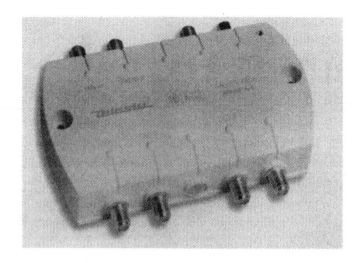

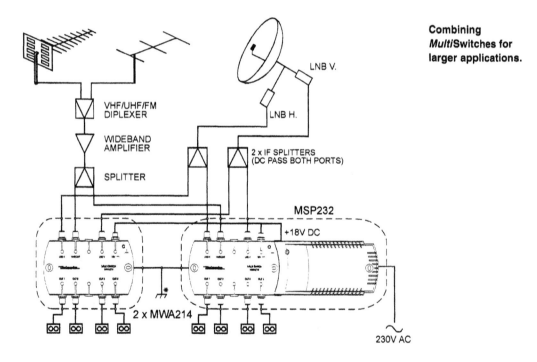

Combining
***Multi*Switches for**
larger applications.

LNB V.

VHF/UHF/FM
DIPLEXER

LNB H.

WIDEBAND
AMPLIFIER

2 x IF SPLITTERS
(DC PASS BOTH PORTS)

SPLITTER

MSP232

+18V DC

2 x MWA214

230V AC

Figure 5.1 *A contemporary i.f. switch. This example also incorporates u.h.f./v.h.f. The diagram illustrates how more than one multi-switch can be combined in a system requiring more outlets. 8-way switches are also widely available (Teleste)*

Safety

When distributing signals we are connecting separate properties such as flats back to a common point and thus if a fault developed in one property that gave rise to a dangerous voltage on the connection to it, that would be present in all other properties unless we provide point to point isolation. Thus this is a requirement of all distribution systems. Isolated outlets and other ways of overcoming the requirement for d.c. control are available from equipment suppliers (see Figure 5.4).

Depolarisation control

One answer to this possible problem is to limit the system to receiving only one polarity. One can then fit a standard LNB with feedhorn directly attached and oriented for the required polarity. I.f. out can then be splitter/amplified and distributed or translated. This may raise objections from those wanting other channels!

To achieve both polarities will require two LNBs although this may be achieved with a single unit containing two LNBs. Traditionally this was done by bolting a pair of units in opposite planes to the dish boom with an *ortho-mode transducer*. This is a passive device which is basically a waveguide splitter allowing waves to pass to both LNBs. No depolariser is involved but the two LNB outputs clearly need to be switched. The Marconi type LNB is available in two twin formats. A device with one output fixed at V and one at H will also require external switching but can then be used to feed large numbers of receivers. A switching twin output unit has two ports again but both will switch to H or V in the traditional voltage switching manner. This represents the best way of providing two receivers (only) with signals.

The switching used to control depolarisation has all the receivers on the system connected to its outputs and by their normal H/V switching, they will obtain signals from the appropriate H or V input to the switch. These devices are often referred to as magic switches. All of this technology may be used in any of the distribution systems mentioned (see Figs 5.3 and 5.4).

Problems that may occur here are only developments of those discussed throughout the book for single systems. It is simply a case of identifying them for what they are. If a single point loses one polarity but others are OK, then the problem can only be with the receiver – check for switching voltage – or the port on the switch. If in doubt, swap ports on the switch to prove. If one port fails, the switch will require replacement unless of course you can utilise a spare one. All the same tests can be used to find no signals on one receiver.

If all receivers are affected, then one should begin to consider the LNB(s) or feeds into the switches. By-passing the switch provides an easy test.

I.f. distribution

Having dealt with the polarity issue we need to consider the first i.f. signal. It will have been amplified and conveyed via coax cable to the property (or room). We can introduce all the experiences of Chapter 3 into fault finding problems with individual legs of a system but what does need emphasising here is the problem of mismatches. With often long and certainly variable lengths of cable, running in sometimes adverse areas, the possibility of a mismatch cannot be over-estimated. Due to these factors, the use of a spectrum analyser is largely essential, certainly for the installation and alignment of such systems and ideally for fault finding. The distribution of MAC (with a wideband requirement) or digital satellite signals with a requirement for minimal phase distortion, means that many systems will require large-scale upgrading if they were not planned with foresight and well maintained. Many distribution systems will consist of low grade cable and poor bandwidth/noise level amplifiers and switches. The initial discovery of this may be when an engineer is called to investigate a fault or calls to try to tune up a new service.

Powering systems

Where a magic switch or single distribution amplifier is used in an i.f. system, they will be line powered by the LNB supply voltage. However, this added to the extra consumption of a dual output LNB may well prove too much for certain receivers. Problems can occur with receivers

45

Figure 5.2   *Satellite remodulation distribution hardware. The sophistication of such equipment can be seen – this indicates the calibre of engineer required for repairs! Also shown is the method of rack and cabinet mounting hardware in an industrial installation (e.g. hotel), the servicing position of this system and also the specifications of such hardware (Teleste)*

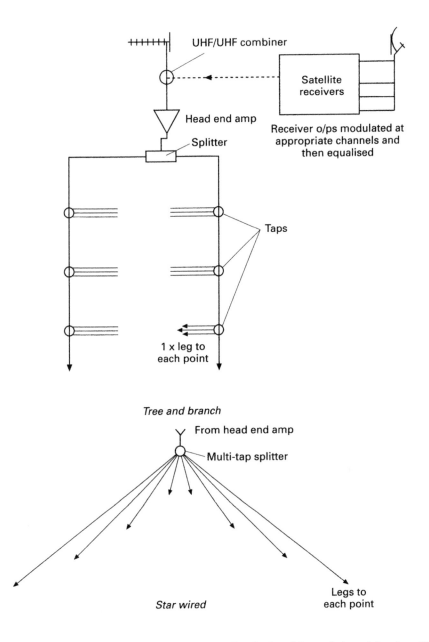

UHF/UHF combiner

Satellite receivers

Head end amp

Splitter

Receiver o/ps modulated at appropriate channels and then equalised

Taps

1 x leg to each point

Tree and branch

From head end amp

Multi-tap splitter

Star wired

Legs to each point

Figure 5.3 *Layout of typical r.f. distribution systems showing headend and how u.h.f. modulated satellite signals are introduced and the cable layouts that could be encountered. The star system splitter has outlets with varying degrees of attenuation akin to the taps on the tree and branch system. Therefore each point is connected directly to this unit making for fewer taps, etc., but more cable running! Fault finding here should be much easier as cable lengths can be simply swapped into another port on the centrally located splitter*

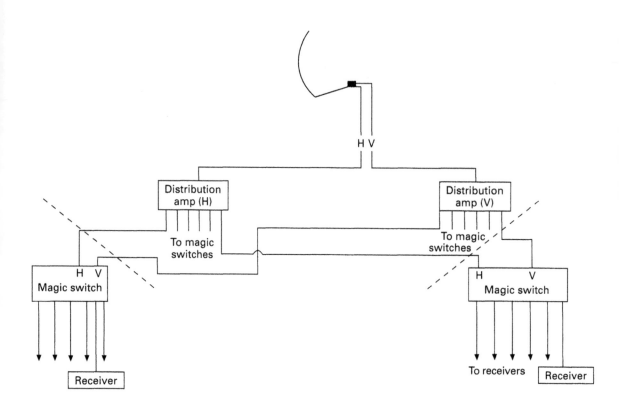

Figure 5.4 *IF distribution system. To achieve point to point isolation, 2 feeds (1 × H, 1 × V) from distribution amps will be fed, via isolating outlets to each property. The magic switch (or a twin input receiver) will then be used in each case.*

intermittently pumping, declaring an 'LNB short' or indeed suffering power supply damage if they are not well protected when too much is powered from the LNB supply. Other problems include chattering of the relays in magic switches. The answer is to externally power these units (which usually involves a different amplifier to a line fed one and so replacement rather than reconnection may have to be the answer). It is possible to introduce a diode fed d.c. supply using a purpose designed d.c. injector unit. U.h.f. amplifiers will always be mains powered (albeit sometimes via a separate power supply unit) and so this is a prime

suspect if a u.h.f. distribution system fails. The d.c. outputs from external PSUs are usually fused and so this is an immediate check. Failures are often due to intermittent cable or connector short circuits where d.c. passes. If you have problems with strange intermittent symptoms and a distribution system is involved, take a good look at the powering methods.

Technical specifications	MHU 244 Hyperband DSB	MHU 245 UHF DSB	Remarks
GENERAL			
Supply voltage, nominal (V)	18	18	15 ... 25V
Current consumption, typically (mA)	600	600	
Operating temperature (°C)	−10 ... + 55	−10 ... +55	
Shielding attenuation (dB)	70	70	30 ... 950 MHz
Radiation (dBpW)	20	20	30 ... 950 MHz
Connectors	F, RCA	F, RCA	
Dimensions (mm)	260 x 190 x 105	260 x 190 x 105	
FM RADIO			
Frequency range (MHz)	87.5 ... 108	87.5 ... 108	50 kHz tuning steps
Input impedance (ohm)	75	75	
Input level range (dBµV)	50 ... 90	50 ... 90	
Harmonic distortion (%)	1	1	typ. 0.4%, dev. 27 kHz
S/N CCIR 567–2 (dB)	47	47	typ. 54 dB, dev. 50 kHz (FM + mod.)
TV MODULATOR			
Frequency range (MHz)	300 ... 470	470 ... 862	8 MHz raster, PAL B/G, PAL I
Frequency tolerance (kHz)	±30	±30	
Output level, nominal (dBµV)	75	751	Note 1
Output level stability (dB)	+1.5	±1.5	
Spurious attenuation (dBc)	60	60	
Video			
Input level (Vpp)	1	1	±3 dB
Input impedance (ohm)	75	75	
Frequency range (Hz/MHz)	25 ... 4.5	25 ... 4.5	
Frquency response (dB)	±2	±2	< −30 dB, ≥5,5 MHz
Group delay (ns)	120	120	
Diff. gain, typ. (%)	10	10	
Diff. phase, typ. (°)	12	12	
S/N CCIR 567–2 (dB)	47	47	
External audio			
Input level (Vpp)	1	1	±6 dB
Input impedance (kohm)	47	47	
Frequency range (Hz)	40 ... 15000	40 ... 15000	
Frequency response (dB)	±2	±2	
Harmonic distortion (%)	1	1	typ. 0.4%, dev. 27 kHz
Pre-emphasis (µs)	50	50	
Audio carrier frequency, (MHz)	5.5/6.0/6.5	5.5/6.0/6.5	progammable
Frequency tolerance (kHz)	+50	+50	
Audio peak deviation (kHz)	±50	±50	
Audio carrier level (dB)	−13	−13	adjustable −10 ... −20 dB
S/N CCIR 468–3 (dB)	47	47	typ. 54 dB, dev. 50 kHz

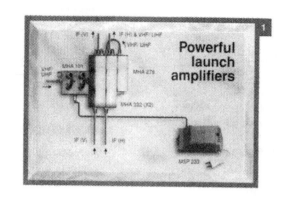

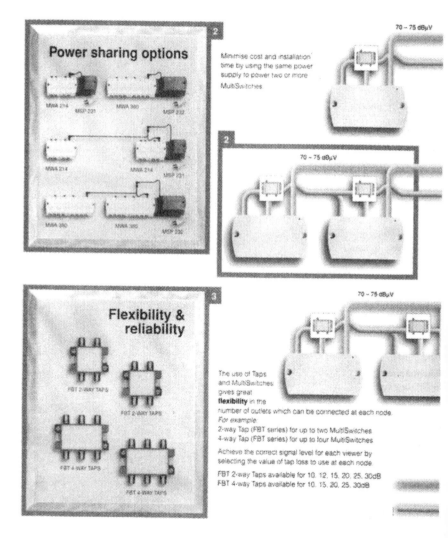

Figure 5.5 *A tree-branch i.f., u.h.f./f.m. distribution system highlighting the hardware used for each task (Teleste)*

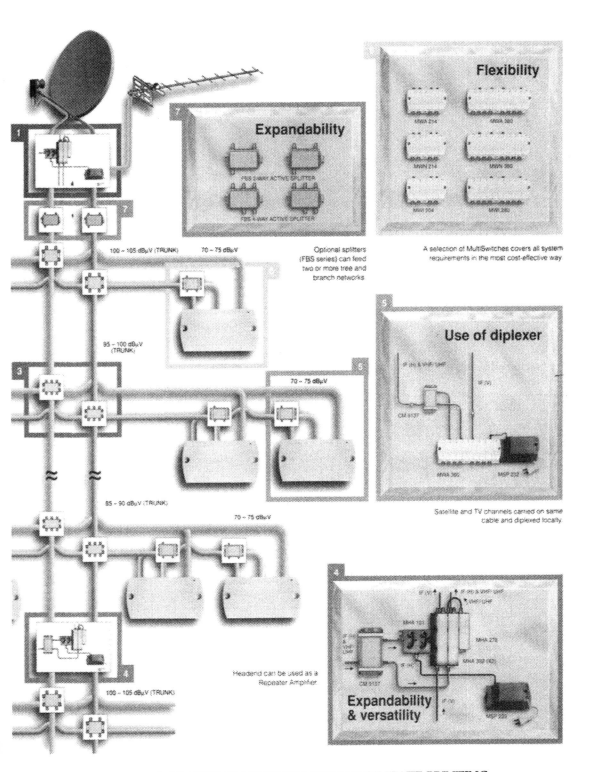

Expandability

FBS 2-WAY ACTIVE SPLITTER

FBS 4-WAY ACTIVE SPLITTER

Optional splitters (FBS series) can feed two or more tree and branch networks

Flexibility

MWA 214

MWA 380

MWN 214

MWN 380

MWI 204

MWI 280

A selection of MultiSwitches covers all system requirements in the most cost-effective way

Use of diplexer

IF (H) & VHF/UHF

IF (V)

CM 9137

MWA 380

MSP 232

Satellite and TV channels carried on same cable and diplexed locally.

Expandability & versatility

IF (V)

IF (H) & VHF/UHF

VHF/UHF

MHA 101

MHA 278

MHA 302 (X2)

IF (H)

CM 9137

IF (V)

MSP 232

Headend can be used as a Repeater Amplifier

100 ~ 105 dBμV (TRUNK)

70 ~ 75 dBμV

95 ~ 100 dBμV (TRUNK)

70 ~ 75 dBμV

85 ~ 90 dBμV (TRUNK)

70 ~ 75 dBμV

100 ~ 105 dBμV (TRUNK)

THIS CIRCUIT HAS BEEN SPLIT TO FACILITATE PRINTING

6

ACTUATORS AND POSITIONERS

Motorised dish systems account for only a tiny fraction of DTH systems in any given area, especially in the UK. However, when problems occur, there is a lot more to consider. In Chapter 3 we discussed the added complexity of dish alignment and mounts. Here we look at how the dish is moved and how that movement is controlled.

This chapter obviously covers detail that is irrelevant to fixed dish systems. Let's define our terms first. The actuator is the motor and gear assembly that is fitted to the dish and mount to drive it from east to west and vice versa. Alternative names include *jack arm* or *screw jack*. The positioner is the electronic drive and control unit that sits inside the house and determines what the actuator does. The positioner may be a separate box external to the receiver, controlled by an interconnecting lead or simply entirely separate or indeed an internal, integral part of the receiver itself.

Figure 6.1 illustrates the construction of a motorised dish system. Virtually all positioners will have the facility to store preset dish positions to enable satellite positions to be quickly and easily reached. Where there is integration between receiver and positioner, the facility will probably exist to preset dish position for each channel preset. Thus the action of simply changing channels will cause necessary movement of the dish position.

To enable this kind of feature, the system needs to know where the dish is to start with and how far it needs to go to reach the desired position. As we can see from Fig. 6.1, we therefore have tacho feedback from the motor/arm to the positioner. Calculations can then be made as to where a dish is and compared with where it needs to be thus enabling drive to the motor. Another vitally important role is played by the feedback – safety.

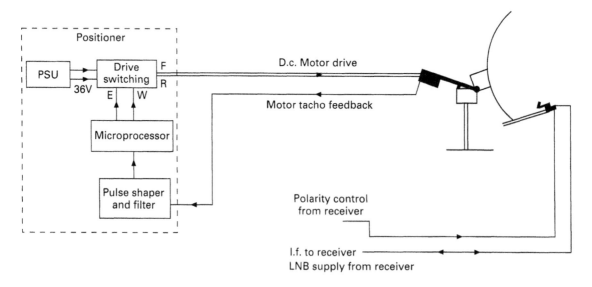

Figure 6.1   *System diagram of a positioner and actuator*

If no feedback is attained when drive has been present to the motor, the positioner will cut off the drive. The obvious assumption is that if there is no feedback, there is no movement and thus continuing to drive the motor is dangerous.

For example, the dish may have been jammed by an obstruction – a tree or a person! Continuing to drive against this obstruction would be likely to lead to damage to the positioner and the motor of the actuator not to mention the obstruction.

The actuator arm

The meat end of the operation is available in a variety of sizes generally expressed as length of arm. Reference to manufacturer's data should be made as to what size of arm should be used with which dish, but typically a 12-inch arm will be used up to a 1 m dish and an 18-inch arm for up to 1.5 m dishes. Arms can be driven over a relatively wide range of voltage but typically this will

be 36 V. The device (Fig 6.2) consists of a motor, mounted at one end suspended from the dish mount. The motor should be mounted upward. The unit will invariably be labelled with a warning and arrow to reinforce this. This is to prevent problems with ingress of water, which is the most common cause of failure with actuators. Gaskets and glands are provided but often get left out at installation. Figure 6.3 shows an example where a cable entry gland is missing and the hole has been sealed, as has the joint in the casings with a silicon rubber compound.

Figure 6.3 *Close-up of motor end of actuator illustrating mounting of motor upward. The warning label can be seen beneath*

The motor drives a worm shaft which interfaces with the dish bracket as a smooth sliding ram with a ball joint at one end (see Fig. 6.4). This shaft will be kept lubricated by the application of grease which is typically applied via a grease nipple on the body of the arm. Some designs are lubrication free – consult the manufacturer's literature.

Actuator feedback

The actuator also consists of some form of device for providing feedback. The most popular idea in modern units is for pulse feedback from either a reed switch or opto device. The principle is the same in that as the motor turns, it invokes either a magnetic or optical effect on the sensor. When using a reed switch, consider that the motor is stopped and the switch is open. A magnet

Figure 6.2 *An actuator arm*

Figure 6.4   *The ram and ball joints of the actuator arm*

attached to the motor gearing turns as the motor turns. As it comes into close proximity to the reed switch, the switch closes and as soon as the motor moves around again, the switch once again opens. Thus a voltage applied to one side of the switch will be pulsed at the other side as long as the motor keeps turning.

With an opto device, a flag turns which interrupts an infra-red beam and thus turns a photo transistor on or off providing pulses. Reed switches fail going either short or open. This is easily proven with an ohmmeter and a magnet! Most designs allows for easy replacement – some are plugged!

The alternative design, now virtually obsolete and not supported by many positioner designs, was a potentiometer feedback where resistance was varied as the motor moved.

Limit switches

Arm designs vary but they will always have at least one end limit switch if not two. The idea is that if the arm reaches the end of its travel inwards or outwards, the motor supply should be cut. By adding diodes, we can ensure that the drive, if reversed, will still allow the dish to be moved the other way. Calibration of the home limit switch is vital to prevent damage to the arm. Most positioners use this point to calibrate the start of the arc or tacho count as 0000. Always check, therefore, that the limit switch is set as per manufacturer's guidelines – this usually involves a mark engraved in the metal casing of the connection box. Loosen the securing screw to adjust the switch and then retighten. If a second limit switch is present for the opposite end of the travel, this can be set to either the mechanical limit of the dish or arm movement or just past the furthest required satellite. Most positioners will supersede this electronically in that they will not drive beyond the stored furthest satellite position. Clearly if limit switches fail to do their job, the results could be quite expensive. Best to replace the whole arm in this instance.

Other actuator problems

Most problems associated with moveable dishes are the failure to move or to reliably move between stored positions. Theoretically there are possibilities of positioner faults but realistically the problems are going to be with the dish mount or actuator. Check for too much movement in the assembly when it is stationary. If it is free to move and settle in a position within a few degrees no wonder it doesn't find the right position. Loose bolts or a broken actuator are likely.

If everything seems firm, there may be too much friction in the system. Unhook one end of the actuator and move the dish by hand. Does it move freely? If so look at the ball joints on the actuator. Can it move (twist) as its attitude to the dish changes through the arc. Corroded or overtightened ball joints are quite common. Strip, clean and relubricate. Then ensure that there is no water in the feedback cable. Scope the pulses at the positioner end of the cable to ensure that they are not disappearing or fluctuating wildly. If they are and motor action seems stable, try a new tacho sensor. If all else fails, suspect the actuator motor itself. They become lazy after some use. The problem can be very intermittent and becomes highly dispiriting.

Another problem occurs if the dish is not moved very often. When it is moved after a period of being in one place, the arm grease may have become stiff or the motor stuck or the arm dirtied by debris. All this can lead to a failure to move (or move quickly enough to provide adequate tacho pulses) or to find its positions correctly. Clean, lubricate and reset to cure. Advise the customer to exercise the dish more often!

Tracing loss of tacho feedback

If the positioner stops driving the dish due to loss of feedback, it can do so very quickly, thus giving little or no time to check for pulses. The answer is to power the dish via an external bench supply whilst 'scoping for the pulses. If they are missing, simple 'scope or continuity tests will determine where. If not, then it is likely that you have a positioner problem. Although motors may normally run from a 36 V supply, using a lower supply does no harm for tests as it means you are less likely to damage anything and by the motor turning more

slowly, you have more time to observe what's going on.

Weather

We have to be aware that these devices are outside in all weathers. When the weather is cold, the d.c. resistance of the motor windings will drop and thus present a heavier load to the drive circuit. You may well experience the positioner cutting out in such conditions. Similarly, if there is heavy wind buffeting the dish as it attempts to move. Many positioners incorporate a thermal fuse to prevent damage and this may also operate if the dish is being repeatedly moved for long periods – maybe as a positioner is being set up – beware!

Actuator replacement

It is likely that you will have to reset all stored dish positions as the tacho feedback will vary from arm to arm.

Positioners

Figure 6.5 shows a circuit diagram of a popular, current dish positioner unit. Although a separate box, it is designed to interface with Pace receivers via pin 12 of the scart socket – we can see this as a data input to the microprocessor.

Parts of this circuit have no significance to it being a positioner – the microprocessor handles via U3 a display drive, indicating dish position, it also has keyscan for on-board controls. A bus between U1 the microprocessor and memory IC U5 handles data storage. An infra-red remote receiver is also present. Of significance to us is the feedback circuit based around U4D which shapes, inverts and buffers the pulses and feeds them into the micro at pin 12. Inverted pulses are also fed in via Q6 and pin 30.

Pulses are counted within the microprocessor and compared against the internal value from memory. If a feedback problem were found to be within the positioner, then it would likely be in the feed from input to microprocessor unit. Simple 'scope signal tracing would show this.

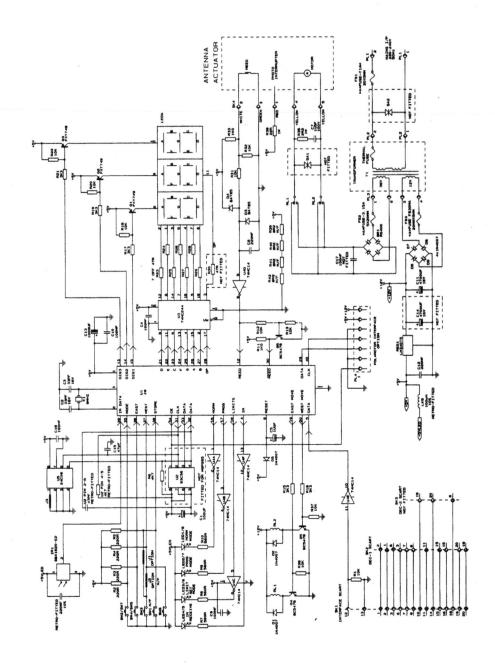

Figure 6.5   *Circuit diagram of modern dish positioner (Pace)*

Motor supply

The power supply for the motor can be seen to be very simple. Here the drive is switched by a pair of relays. Alternative arrangements may use thyristor or triac switching. The drive for the switching will be from the microprocessor, here at pins 19 and 20 via coil drivers Q4 and Q5. Failures here are likely to be caused by problems with the motor. However, if a failure were to develop here first – e.g. Q4 went short circuit collector to emitter and there was no limit switch protection on the arm – in many designs it is likely that the motor would succumb before the power supply fuse went.

Naturally any power handling circuit is prone to failure.

Initialisation

As already stated, the positioner needs a reference to know where point zero is. This is usually taken to be the home limit switch. In other words the dish is driven home until it is stopped by the limit switch. Pulse counts will then start from here. When a unit is powered after being off completely, many designs will drive the dish to this point to recover its home position from where it will then recount to where it wants to be. It is after such an event that many relocation problems are experienced!

Alternative dish positioners

IRTE, which also have an LNB positioning device called the Multisat (discussed in Chapter 3), have produced a dish actuator/positioner (the Omni Sat) suitable for dishes up to 1 m in diameter which controls movement both horizontally and vertically, thus removing the need for accurate dish mount alignment. It also has the benefit of controlling all this via the single coaxial cable used for a fixed dish thus enabling rapid conversions to be motorised.

7

TUNERS AND TUNING SYSTEMS

There are very distinct parallels between the tuning systems used in satellite receivers and those encountered by many readers in TV and VCR. Conversely there are a number of interesting differences. We shall see throughout this chapter what can go wrong and what to do about it. There are many theoretical factors relating to tuning and tuners that should be absorbed from other sources.

The tuner

In Fig. 7.1, we can see the typical extent of a satellite receiver tuner which is a self-contained can module as with most varicap TV and VCR tuners. MOD1, as it is referenced in this example, takes the first i.f. input (950–1750 MHz in the case of this non-1D receiver), tunes to the required carrier and then down-converts to an i.f. of 479.5

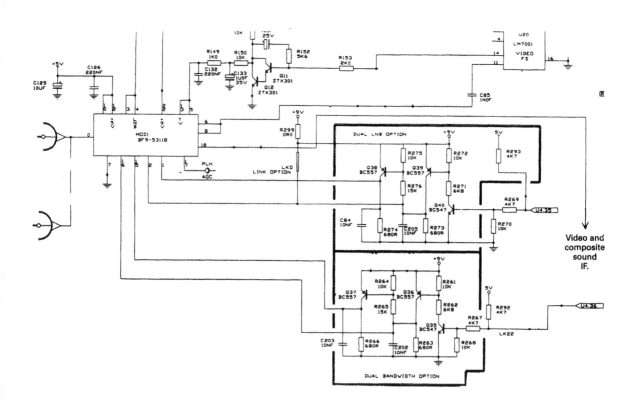

Figure 7.1 *Tuner circuitry from an early IRD. Illustrates the ins and outs of the tuner module (Pace)*

MHz, which has a 27 MHz bandwidth in receivers designed specifically for Astra and wider, 36 MHz for most other satellites. Our example can be seen to have the option to be able to switch bandwidths.

The example shown allows the option of a pair of dish inputs and the necessary switching to achieve this.

The tuning is a d.c. voltage applied to pins 13 and 5 in the standard mould of varicap tuning. This would be referred to as BT in a TV or VCR. With this earlier receiver, the voltage swings between 0 and 24 V. In later, wider input bandwidth designs, it has a wider range to 33 V or so. Our design can be seen to be frequency synthesis, a system described below.

Video demodulation

Still within the tuner unit, the vision is demodulated (PLL demodulator) and output, with the composite sound i.f. at pin 18. As stated in the next chapter, we can 'scope video at this point but it is very weak and noisy.

Tuner faults and repairs

Generally speaking it is not possible in a conventional servicing workshop to fault find within or carry out repairs to the tuner module. The standard practice is to replace the faulty module or have it rebuilt by a specialist such as MCES (see Appendix 3). The cost of a replacement tuner is usually relatively high compared with the cost of a replacement receiver and so where the tuners fail, it is often not viable to carry out the repair. All of these factors make it very important that diagnosis of a faulty tuner is correct.

Certain problems can sometimes be cured by simple repairs to the tuner units – these will be highlighted as we discuss failures. Never attempt to repair a tuner unless you can afford to wreck it!

If the problem is that there are no signals and the receiver is definitely the cause of the problems, then we first need to look at some obvious points. We are assuming that there is noise (snow) on screen and noise (hiss) on sound (when any muting is removed). Is there a supply to the LNB? This can be measured at the first i.f. connector without even removing the lid of the receiver. If not, ensure that it is present to the designated pin

of the tuner – here pins 3 and 4. If it is, then it is being lost in the tuner and the cause ought to be simple. Examine the tuner for signs of dry joints, especially on the input connector. Following this, ensure that the d.c. supplies for the tuner are present – 5 V on pins 10 and 12 and 9 V on pin 22. Ensure also that grounds are intact especially the case lugs of the tuner. Due to it being a large lump of metal, the tuner is often used as a means of connecting various lands of print together and to ground and so drys on its case lugs can cause many odd symptoms. Next check that the tuning supply, pin 5 and 13 in Fig. 7.1, is present and varies as the tuning is moved – e.g. when changing channel. If it is stuck at one point (usually top or bottom of voltage range), then you are likely to have a problem with the tuning circuit rather than the tuner. If any voltage on a tuner pin reads very low or zero, check that there is not a short circuit to ground inside the tuner. Having made these checks on the type of tuner circuit shown, it is likely that the tuner is faulty.

If the tuning is carried out within the tuner unit and the only external influence is via a data bus such as in Fig. 7.2, then having checked for activity on the bus (see Chapter 11), one can only really conclude that the tuner is faulty.

Do make sure that you don't get caught out by receivers with more than one dish input and being plugged into the wrong one!

Problems with demodulated vision traced to within the tuner's demodulator section (as opposed to the remainder of the video path) may be repaired without replacing the tuner. If the fault is along the lines of corrupted sync or hummy vision or negative vision, look inside the tuner for electrolytic capacitors. Replace any in the demodulator or BT area and see if this cures the problem. Clues as to the likelihood of this would be that the problem improves as the unit warms up. Scoping the video out of the tuner may be sufficient to prove that the fault is in the tuner but it may be necessary to inject signals through the rest of the videopath to exonerate it.

A low signals fault proven to be in the receiver is almost certainly going to be caused by the tuner and require its replacement. Confirm that all supplies are present and clean. Eliminate any AGC control by replacing it with a diode isolated external supply. If the results are still low, then the tuner is in trouble. It may be possible to resolder joints in the r.f. stages of the tuner to cure such a problem but device failure is more likely.

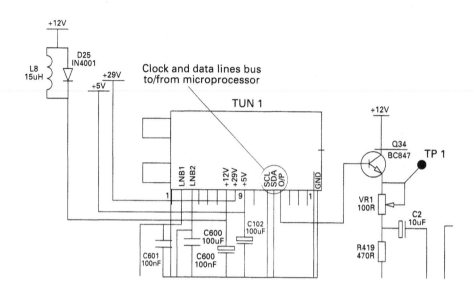

Figure 7.2   *Tuner circuit where tuning is internal and invoked via the I₂C bus (Grove Farm Publications)*

In their normal working state, tuners can be very microphonic – this means that tapping them will give rise to noise or line interference on vision. Only experience can differentiate between normal levels of this and a fault. Where such a fault exists, it may again be possible to cure by removing the side cases of the tuner, resoldering particularly ground joints within the tuner and then tightening the side retaining clips and tensioners and refitting the sides.

Tuning systems

Invariably, tuning is based on a frequency synthesis system (see Fig. 7.3). The frequency synthesising IC is instructed by the system control (tuning microprocessor if separate) circuit, which frequency is required. More specifically, its internal programmable counter derived from its external reference is programmed with the number which equates to the divided down local oscillator or prescaler from the tuner. The synthesising circuit then varies the tuning voltage until the feedback from the tuner's local oscillator and its counter coincide. In the diagram, the local oscillator is divided by 128 to provide the prescaler output (Fig. 7.4).

The tuning voltage is controlled by the synthesiser IC but it does not output a varying d.c. Dependent on design, the ouput may be pulse width or pulse density modulated. This drives an integrator/low pass filter circuit which taps BT from a supply of 24 or 33 V. All of this is illustrated in Fig. 7.5, the same principles may be used for audio and r.f. modulator tuning.

Older or cruder tuning systems may be encountered where the tuning voltage is simply varied by a potentiometer or a series of preset resistors which are then switched in when the channel is selected, i.e. they 'store' a tuning point. This is common to early TV and VCR tuning systems. Others may electronically store information and vary the tuning supply but not use any form of feedback from the tuner to confirm specific frequencies. These systems are termed voltage synthesis. They are simpler and the circuits involved work exactly as those already described minus the feedback and frequency comparison.

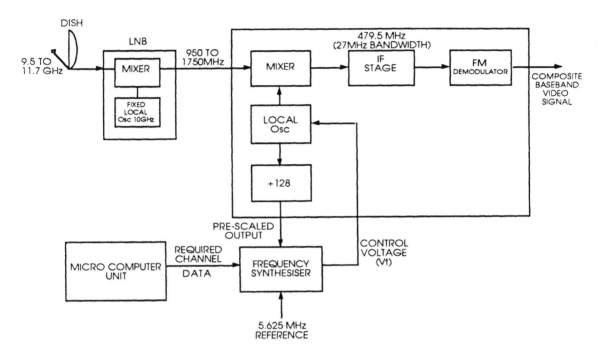

Figure 7.3 *Layout of a frequency synthesis system*

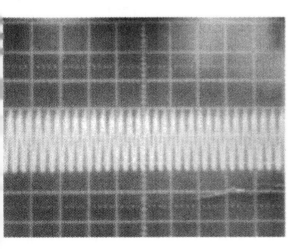

Figure 7.4 *Prescaler out of tuner as viewed on oscilloscope. Frequency will depend on design of circuit*

A search facility is invoked which gradually increases or decreases the tuning volts whilst a sync detection circuit monitors the demodulator output. When sync is detected, a status line triggers the sweep to stop. If the channel is the desired one, the tuning point is stored by the operator, if not the sweep is restarted. Therefore, due to the lack of definitive frequency feedback, if tolerances shift outside of the range of AFC, the tuning point may drift necessitating periodic retuning.

Tuning system fault finding

Firstly let's consider the situation where you have determined the cause of no signals to be a problem with the tuning voltage, BT. If there is no or very low BT and it doesn't seem to vary much, check that the tuning supply from which it is tapped is present. This will be a 24–33 V supply. In a synthesis system, this will be present on the load

61

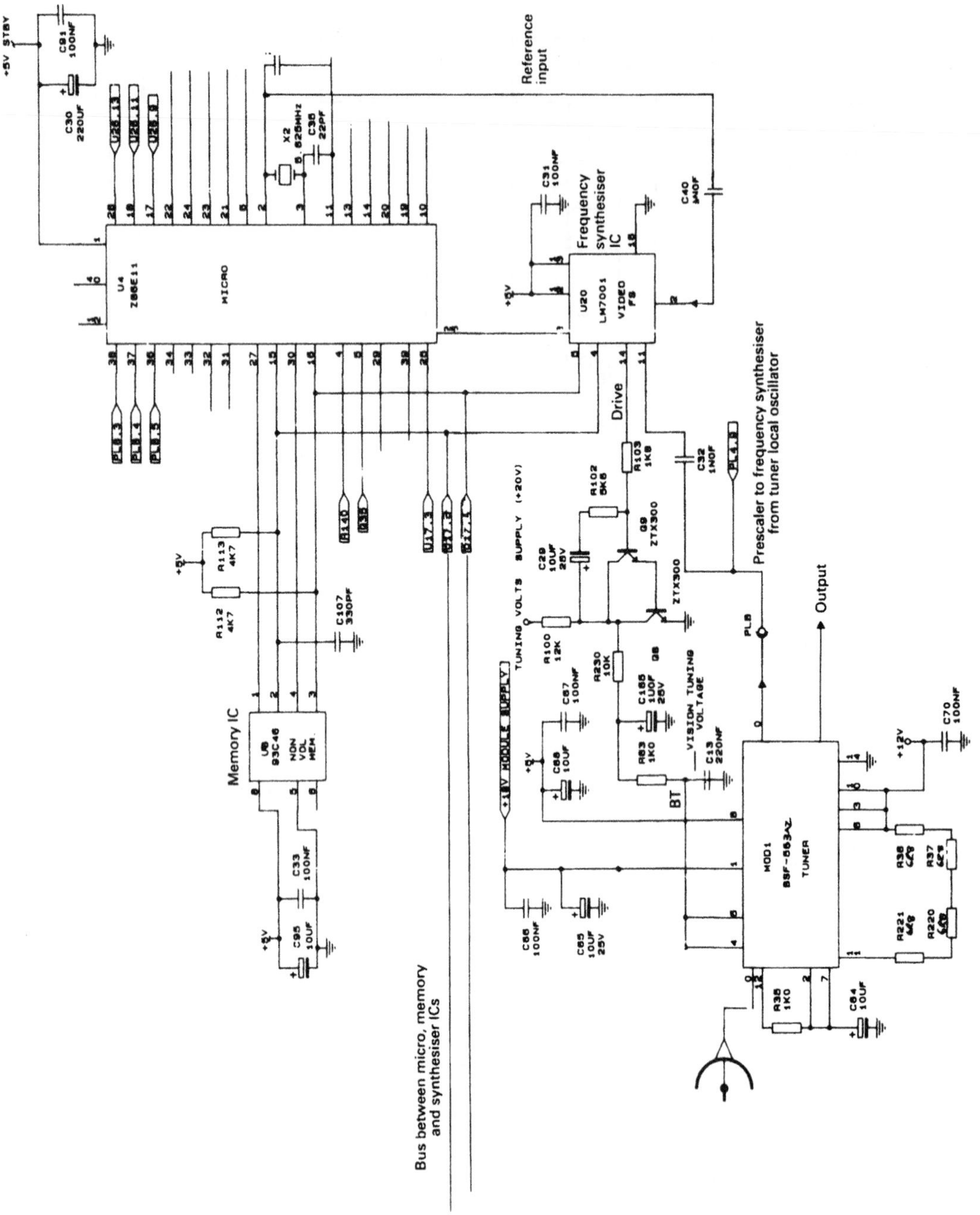

Figure 7.5 Circuit of a frequency synthesis tuning system (Pace)

resistor at collector of the transistor used to develop the BT. On tuners where all tuning is internal, this supply will be present on one pin of the tuner module. A power supply fault, open circuit collector feed resistor or short circuit in the LPF transistor are likely causes.

If this doesn't prove to be the case or if the BT is permanently stuck at the maximum supply voltage (24–33 V) then suspect that there is a problem with the modulated drive from the tuning IC. Set the unit sweep tuning and then 'scoping the output of the IC will in some cases allow you to see the drive varying in pulse width or density. In the latter case, this can be quite difficult to appreciate. What is appreciable is when the IC output is permanently high or low. In this case, the biggest suspect is a loss of prescaler from the tuner. The synthesiser thinks that the frequency is just too low and so drives hard to increase the tuning point to no avail. Ensure that there are no shorts but the likely cause is the tuner. Other problems occur when there is prescaler output but the synthesiser is faulty. It is a fact that if the prescaler is present, the fault is likely to be in the synthesiser IC. Theoretically the prescaler may be of the wrong frequency – where you know the divider factor as in Fig. 7.2, you can measure the prescaler and calculate its accuracy – but this is not often the case.

If the drive is present and apparently correct but BT remains stuck as maximum supply, suspect an open circuit in the LPF/integrator circuit, especially an open circuit transistor.

Tuning drift can be a real problem due to its invariably intermittent nature. It is rarely encountered on frequency synthesis systems. One can appreciate the reasons from the understanding that we now have of how they operate. Where we do have problems, then we must decide whether the cause is in the tuner, the tuning voltage, BT or the AFC. Many tuning systems will not encompass an external AFC thus making the problem easier to fault find. Either way, the best idea is to remove the receiver's BT to the tuner and replace it with a bench supply, diode-isolated for safety. Use the bench supply to tune up a channel and then see if the drift occurs. If it does and the AFC can be similarly vindicated, then the tuner is the cause of the problem. If it no longer drifts, then monitor the receiver's BT and see if that shifts. Varying BT is the likely cause of any drift outside of the tuner. Look particularly for any decoupling capacitor on the BT line and also supply reservoir capacitors on the tuning supply.

Noise or hum on the tuning supply will, as well as tuning drift, give rise to affects on vision. Therefore when looking for vision faults, it is valid to check, BT with a 'scope.

Where AFC problems occur, this will doubtless be due to drift or misalignment of the detector coil or i.f. offset. Refer to the manufacturer's instructions to calibrate these. It has to be said that any i.f. tweaks are best left well alone unless absolutely necessary.

8

VIDEO PROCESSING CIRCUITS

The areas covered by this chapter account for a fair percentage of the content of a satellite receiver. They contain some exacting techniques and processes that could theoretically lead to quite a few nasty faults. In practice, video circuits are very reliable and perhaps the single most common failure is loss of vision. Please remember that many vision faults are caused by power supply problems and so check supplies with an oscilloscope before getting too deeply involved in this area. The area considered extends from the video demodulator through de-emphasis and filtering,

energy dispersal clamping and video routing and switching.

Here we are discussing a PAL receiver. We shall also discuss take-offs for external decoders such as for MAC.

Demodulators

The input to the demodulator is the tuned channel at second i.f., typically 479.5 MHz. The

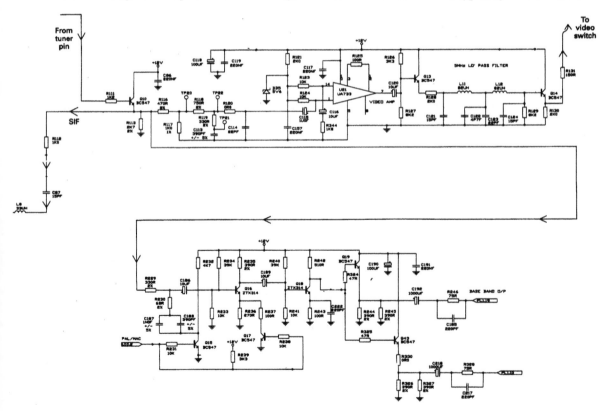

Figure 8.1 *Path of the video signal from demodulator within tuner to input switching. Also shown is baseband amplifier for external decoder output (Pace)*

ision demodulator is usually contained within the tuner unit – unlike the majority of TV/VCR tuners – and thus we must consider that the earliest point at which we can reasonably test for vision is on one pin of the tuner (Fig. 8.1). Here we see demodulated video (and sound i.f.) appearing at pin 18 of the tuner, MOD1. It is buffered by Q10, SIF is filtered off via C87 and the video signal is passed for low pass filtering and clamping but also a path exists, via R229 and C186, for a feed-out to the decoder scart socket. Clearly this can be seen to be prior to de-emphasis and clamping.

Decoder take-off

The lower section of Fig. 8.1 shows the baseband amplifier providing the video output to the decoder socket. De-emphasis can be switched between PAL and MAC, the control from the microprocessor U4 at R231 and the PAL filter formed by R230, C187/8 switched by Q15. When MAC is selected, the amplifier response is flat. Clearly misoperation in the form of setting this de-emphasis option incorrectly would lead to poor or unreliable operation of any connected decoder.

The one decoder scart socket (PL1) is actually used to enable the connecting of two decoders, hence the extra amplifier stage to pin 10 of PL1. The output of the decoder is fed back into the receiver's video switching circuit at pins 3 and 5 of U7B. This is seen in Fig. 8.2.

Another common error is where the customer connects his TV or VCR to the decoder scart socket (despite them being clearly labelled!). The result is a flickering picture varying in level and generally looking very sick – the video here is of course unclamped.

Video signal

It is very important to understand that the video signal present at the output of the demodulator, and indeed well into the video chain, will not be the archetypal 1 V p–p clean video signal you may expect. It will be low level, noisy and indistinct (Fig. 8.3).

Sparklies

It is very difficult to know where to put this section. Does one put it in with dishes where the cause of most sparklies are dealt with? How about LNBs and depolarisers? It's possible that the tuner may be in trouble so what about in Chapter 7? I have decided to discuss the matter here because this is where we discuss the demodulator and that is the circuit that is responsible for producing sparklies (for whatever reason).

What are sparklies?

Simply they are noise, specifically *impulse* noise. They appear on vision as long duration, comet-shaped dots either black or white in appearance. The effect is usually noticed on vision before the audio due to most people's more critical appraisal of vision than sound, but the effect on sound will be crackling and disturbance or hiss. The theory is well documented elsewhere but we must consider the basics. When the signal level input to the demodulator reaches a level of C/N (carrier to noise) that is too low, the demodulator can no longer distinguish between the signal and the noise. It thus produces its own noise. The point at which this occurs is the receiver's threshold. Thus with a marginal C/N ratio from the dish, an older receiver with a threshold of say 8 dB will look more sparkly than one with a 5 dB threshold. These subtleties need to be borne in mind when substituting units for test purposes.

What the sparklies tell us

There is significant relevance in whether you have black, white or both types of sparklies. Similarly whether they are present in roughly equal amounts across all channels from the same satellite (allowing for the natural signal level variations between services). If you have black *and* white sparklies present at the same time and across most channels evenly, then you have weak signals (poor C/N ratio) reaching the demodulator input. I state this most specifically as this means that the cause of the weak signal may be in the receiver prior to the demodulator – i.e. the tuner (r.f. amplifier). It is not always a problem with the feed to the receiver (although of course it is far more likely to be!). Testing signal level with a signal strength meter can in such circumstances prove inconclusive. Level outputs from LNBs vary and it can be

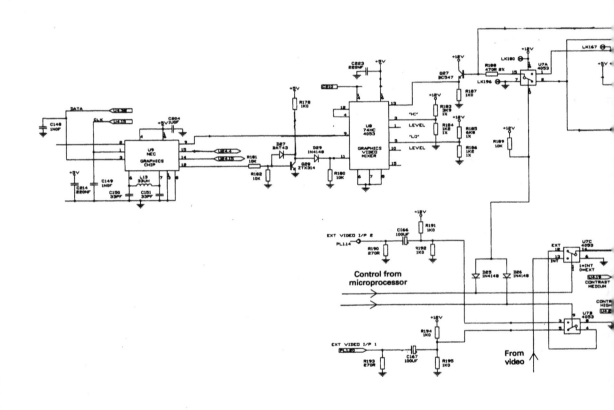

Figure 8.2 Continuation of video path of same receiver. Note: PL7 is the connector for the internal Videocrypt decoder (Pace)

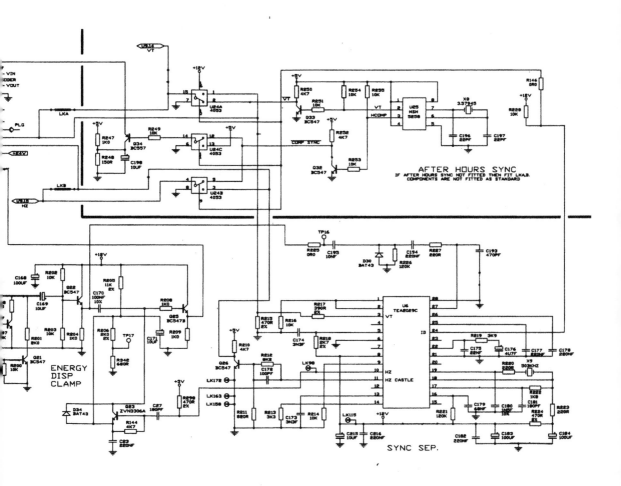

THIS CIRCUIT HAS BEEN SPLIT TO FACILITATE PRINTING

Figure 8.3   *An oscillogram showing the likely state of the video signal out of the demodulator viewed at field rate*

receiver (see Chapter 7). If not, then refer to Chapter 3, 'Weak signals'.

If you have only black sparklies, it indicates that the tuning point is set too low – i.e. below the carrier. White sparklies indicate that the tuning point is above the carrier. If this occurs on only a couple of channels, it indicates simple mistuning especially if one channel has black sparklies only and the other ones have white ones. Reset the tuning point to the exact carrier frequency for the channel. If there is an effect across most channels and the sparklies are consistently either black or white, then this again indicates that the tuning point is low or high but it would not be sensible to reset all tuning points to compensate. You would almost certainly find that they are reading correctly anyway. This overall shift indicates a problem with the i.f. offset (see Chapter 4) or an AFC problem (see Chapter 7).

difficult to discern what part of the brand you are metering with many meters. However, it will highlight a clearly very low signal. A spectrum analyser will be able to give you far more detailed and graphical information about the problem. Probably the simplest way of dealing with the issue in the first instance is to try a replacement receiver. If that looks fine, then you have a suspect

Things to watch out for

The tendency is to overlook the fact that vision from satellite is frequency modulated. That, after all, is why we get the sparkly type of noise. The result of this is that because the chroma subcarrier sits in the thick end of the wedge shaped noise spectrum of an f.m. PAL signal, its carrier to noise ratio (C/N) will be much worse than that of the

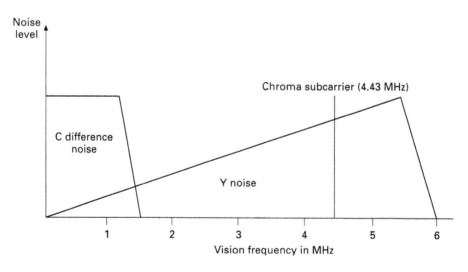

Figure 8.4   *Noise spectrum of a frequency modulated PAL video signal*

luminance (see Fig. 8.4). This is why we still see sparklies (black and white) in deeply saturated chroma even when we have apparently very good signals. With the size of dish, standard of LNB and receiver used for a typical Astra installation, you are not going to eliminate this noise. Do not be misled into thinking that you have some kind of fault when seeing this problem. Ultimately, using large dishes and low noise LNBs with very good threshold receivers will improve matters (although possibly cause others!). But we have to strike a practical balance. Similarly if you want an indication of how clean your signals are, then look at some deeply saturated chroma on a PAL transmission.

Video de-emphasis, amplifiers and filters

De-emphasis is applied to the demodulated vision after a take-off for any decoder interface. In Fig. 8.1, the components concerned are C113/4. Any suggestions for fault finding here are speculative as this area of a receiver is inherently reliable. However, video response problems would be the result – poor h.f. resolution or whiting out are examples. On more specialist receivers there may be facilities to switch between different video de-emphasis standards. This being incorrectly set would be the most likely cause of a problem. The video is next fed to the video amplifier.

Based around transistors or operational amplifiers, the circuitry is simple and reliable. Figure 8.3 illustrates just how simple. U21 is the amplifer, gain is set by R125. Loss of video through this circuit is the only likely problem and as with any circuit, it is a case of signal tracing through the stages. If video reaches the input at pin 1 but there is no output at pin 7, then consider the supply to the IC, the biasing of the output pin via the biasing of pin 14, the inverting input by D35 and potential divider R126/7.

The output at pin 7 is fed to the low pass filtering circuit based around L11/2 and C121/2/3/4. Q13 and Q14 buffer the video in and out of the filter. Again the signal can be 'scoped through all of these stages for fault finding.

In this area of the circuit, a contrast control may be provided as it is in Fig. 8.1, via Q20 and Q21 switching in resistors to ground across the video signal. This is of course totally separate to the

controls on the TV and can be used to balance the satellite video level with that of the terrestrial signal. A word of warning – these controls are prior to the Videocrypt decoder in many cases and so if you reduce the contrast (i.e. video) level too much, the decoder will not operate reliably or at all! This is a very common problem.

Video input switching

As we shall see in the next chapter when considering audio, we have to distribute and route (video) signals at the appropriate time and level to the appropriate places. The principles involved are again very simple – switching is carried out within ICs or by transistors switched by logic levels from system control. Despite this simplicity, the amount of switching involved and the paths followed may become very difficult to keep up with. Process or truth tables may be provided in service manuals to help – i.e. what is fed where in which modes and what logic levels are present to achieve this. The next stage encountered in our circuit (Fig. 8.2) is input switching – i.e. are we routing the video that we have just followed from the tuner/demod or are we to switch in one of the two possible inputs from the decoder socket? This is the purpose of U7B/C.

Energy dispersal clamp

Prior to transmission, a 25 Hz triangular waveform is added to the video signal which thus sits on it. The purpose of this is to prevent continuous peak transmission which can occur in many video scenes and cause interference to other frequencies. This added triangular wave causes the energy to be dispersed by virtue of its constant variation.

We can see that if it were not removed during reception, the viewed picture would flicker at this rate. The process of removing the dispersal waveform is carried out next in the video path and involves clamping our video signal to a fixed reference level during the line sync or burst period. This is carried out in Fig. 8.2 by Q22/3/5 and associated circuitry. Buffered in by Q22 it is clamped to the reference set by R205/6 and R342 by Q23 which is gated by differentiated line sync pulses from the sync separator IC U6. Q25 buffers the signal out.

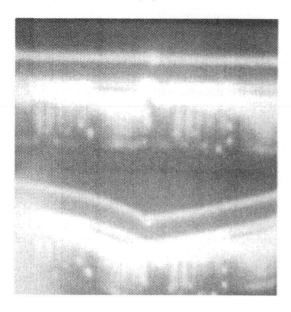

Figure 8.5 *Before (top) and after (bottom) the energy dispersal clamp. The 25 Hz ramp can be clearly seen*

The potential problems here are loss of video altogether through normal problems already discussed, or the failure of the circuit to clamp, thus giving the flickering picture. If the latter occurs, check for line pulses and the action of the clamping transistor. Also check the d.c. reference level (see Fig. 8.5).

Sync separator

This performs the role the same as in TV or VCR. Video is fed to it and it removes line and field sync and burst gating pulses. We have already seen that line sync is required for energy dispersal clamping and so were it to be lost due to a fault here, we would have flickering vision. Burst gating is required for timing within the Videocrypt decoder. H and V sync are needed for graphics insertion and in Fig. 8.2 feeds can be seen from U6 via U24 and U9. It should be noted that some receivers, including the one depicted when the optional 'after hours sync' is not fitted, rely on the video signal being present to lock graphics – i.e. no

signal/picture, then no locked graphics. The alternative with other designs is to have their own sync generator which replaces off-air sync.

Video output switching and on-screen graphics

Further down the path, after energy dispersal clamping, we shall have to switch between feeding the video signal to the outputs directly or via the internal videocrypt decoder. Feeds will have to go to the r.f. modulator and the various scart outlets (see Fig. 8.6). There is a consideration here though: on-screen graphics are an almost standard feature of satellite receivers and so prior to sending the video signal out, these graphics must be inserted. We can see this around ICs U8 and U9 in Fig. 8.2. It is not prudent, however, to have these graphics present on the video feed to the scart connected to a VCR, otherwise they will be recorded and thus should you bring up a menu on the satellite receiver, it would appear on the recording! Therefore, video fed to the VCR scart is taken off before graphics insertion.

On-screen graphics

Again, this is an area of high reliability. Modern units have menus with their own coloured backgrounds. These backgrounds need to be generated using a subcarrier oscillator. It is not unusual to encounter drift in these circuits and thus no colour on graphics (see Fig. 8.7). Attention should be paid to the oscillator around X2 (17.734 MHz = $4 \times$ FSC of 4.43 MHz).

All graphics generation and insertion is controlled by the system control microprocessor, the other governing factors being line and field sync. These then are the areas to check if you have problems with displays. If you have no displays, then obviously check supplies and drives to the generator circuit but also its master clock which determines timing – i.e. position. In Fig. 8.7, this is within U10 and so failure will require replacement of the IC.

Video mutes may well be employed and this is where they are generated. A sync detection circuit will automatically cause a blank, coloured raster to be displayed if there is no sync present. Logically this would be in the presence of no or very weak

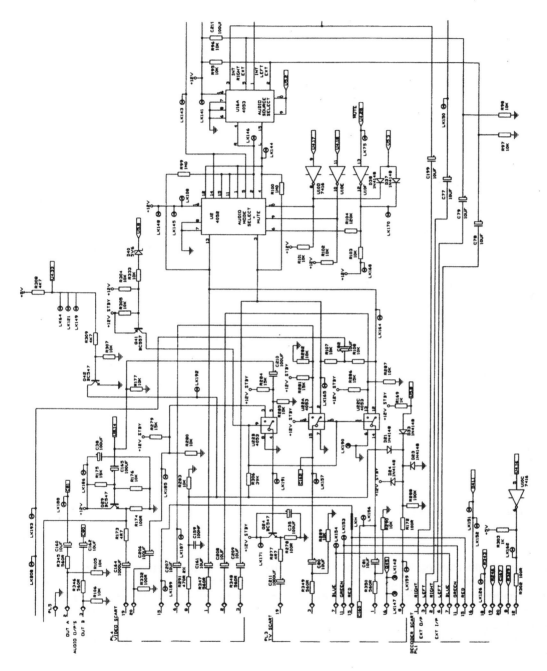

Figure 8.6   *The audio and video switching arrangements to provide correct outputs to TV and VCR at scart (Pace)*

71

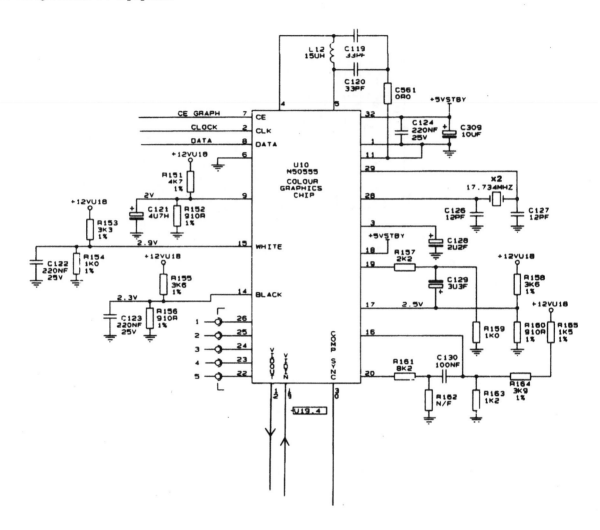

Figure 8.7   *A colour graphics/display generator circuit (Pace)*

signal. However, faults causing a loss of sync here would result in this blank raster. Some receivers enable this facility to be switched off and this should be one of the first moves you make in fault finding – see what is truly present. If this is not an option, simply overriding a status line within system control should enable the same result.

Present designs

Throughout the descriptions in this chapter, examples have been used from older circuit designs. This is because they are more discrete and thus enable us to see the construction, layout and process of a path. The more modern designs are far more integrated and at best video signals will loop in and out of an IC. Figure 8.8 illustrates this modern trend – here the audio is also processed within the one IC.

Teletext

The majority of satellite TV broadcasts contain teletext services. There are exceptions – some of

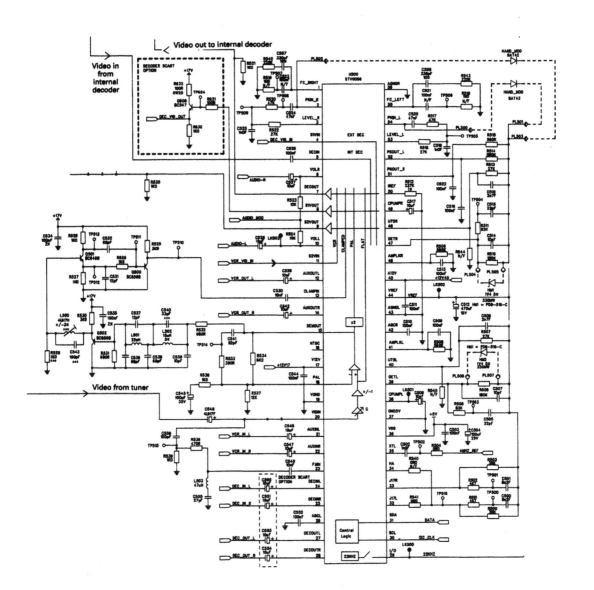

Figure 8.8   *Modern A/V processing circuit (Pace)*

the German services via Astra do not. With PAL transmissions, the teletext decoder in the television is perfectly capable of decoding the data and so the same TV handset is used in the normal way to view the text. However, certain designs of TV will not work with signals fed in via scart – if you encounter a problem, check via r.f. first. MAC services containing teletext will have the text processed by the MAC decoder (often an external box) and so not use the TV handset.

Problems with teletext often manifest themselves as corruption. This is where a page of text has missing or incorrect characters. The degree will reflect the severity of the problem. Whilst it is possible that problems in the receiver could cause text problems, virtually all such faults are caused by a signal fault feeding the receiver. The obvious potential problem is that of insufficient signal. All the procedures discussed to date should be implemented to ensure that the signal level is at its maximum. Sparklies on screen are likely to mean text corruption. One must never assume that weak signals are always the cause of such problems, however – the mismatch is back! Reflections in the cable will cause absolute mayhem in a teletext system and so earlier advice on mismatches could well be needed to find a teletext fault.

Never overlook the calibre of text decoder in the television. Older designs and even some newer ones do not handle data from weak signals well. Have a look at performance on terrestrial signals as a comparison. Try a replacement TV as a test if you begin to wonder about the one in use.

In a receiver, response problems in the video path will cause text problems. If the de-emphasis can be switched, ensure that it is correctly set for each channel. Video level may be adjustable and if too low will cause poor text decoding. Teletext can be used in certain circumstances as an installation or fault finding aid – it is unaffected by Videocrypt scrambling and so can be used as a guide to video signal quality where you have a descrambling problem. If text is clean, it is likely that you have a faulty Videocrypt decoder; where the text is severely corrupt and the Videocrypt appears not to work, you probably have a fault with the video signal fed to it.

9

RECEIVER AUDIO CIRCUITS

Unlike a traditional, terrestrial TV receiver, the audio circuits of a typical satellite receiver are rather complex. They are invariably stereo (although mono systems still exist, particularly in SMATV form), with the possibility of variable i.f. bandwidth and switchable de-emphasis time constants among others.

Having said this, the results are usually that the symptoms are much the same as traditional TV but you have much more circuitry to search through to find the cause. As we will see, much audio processing circuitry in older receivers is based around simple op amps and so much signal tracing in and out of ICs is the order of the day and later units, certainly in the mainstream, Astra market where designs are cost conscious, tend to work in the digital domain.

A brief overview of satellite TV sound transmission seems prudent. Details between satellites will vary and some sample data is included in Fig. 9.1. All sound is frequency modulated onto a carrier sitting, initially around 5.5 MHz or higher above the vision carrier – no great difference to terrestrial there then. However, this is only the initial or

Television channels indicated in Bold type, Radio indicated in Italics

XP	FREQ.	BEAM	CHANNEL	VIDEO	ENCRYPT	HOURS	AUDIO FREQ.	LANGUAGE	TEXT
EUTELSAT II-F2 (10 Degrees East)									
25V	10.972	Wide	**VTV Cable TV**	PAL	None	18	6.60, 7.20	Slovak	Y
20H	10.987	Wide	**ATV**	PAL	None	19	6.65	Turkish	Y
	10.987	Wide	*Kiss FM*				7.20	*Turkish*	
	10.987	Wide	*Radiosport*				7.56	*Turkish*	
20H	11.017	Wide	**Satel2 (test)**	PAL	None		6.65	Turkish	
21H	11.080	Wide	**Europe by Satellite**	PAL	None	10	6.60, 7.02, 7.20, 7.38, 7.56	raw, raw, GB, F, D	Y
26V	11.095	Wide	**TGRT-TV**	PAL	None		6.65	Turkish	
	11.095	Wide	*TGRT-FM*				7.38, 8.02	*Turkish*	
37V	11.163	Wide	**Worldnet/C-SPAN §**	PAL	None	6 M-F	6.60	English/Arabic	
	11.575	Wide	*VoA and Worldnet (audio feed)*			6 M-F	7.38	*English*	
	11.575	Wide	*Voice of America (Europe)*			6 M-F	7.56	*English*	
	11.575	Wide	**MED-TV**	PAL	None	3	6.60, 7.20	Kurdish	
33H	11.596	Wide	**ET1**	PAL	None	17	6.60	Greek	
	11.596	Wide	*ERA2*				6.60*	*Greek*	
38V	11.617	Wide	**InterStar**	PAL	None	24	6.65	Turkish	
	11.617	Wide	*Metro FM*				7.02/7.20	*Turkish*	
	11.617	Wide	*Kral FM*				7.38/7.56	*Turkish*	
	11.617	Wide	*Radio Maria*				7.74	*Italian*	
	11.617	Wide	*Super FM*				8.10/8.28	*Turkish*	
39V	11.658	Wide	**RTP Internacional**	PAL	None	24	6.60	Portuguese	
	11.658	Wide	*RDP Internacional*				7.02	*Portuguese*	
	11.658	Wide	*RDP Antena 1*				7.20	*Portuguese*	
	11.658	Wide	*Rádio Renascença 1*				7.38/7.56	*Portuguese*	
	11.658	Wide	*RFM (Radio Renascença 2)*				7.74/7.92	*Portuguese*	
	11.658	Wide	*Rádio Comercial*				8.45	*Portuguese*	

Figure 9.1   *Audio carrier detail for a given satellite's channels (Eutelsat)*

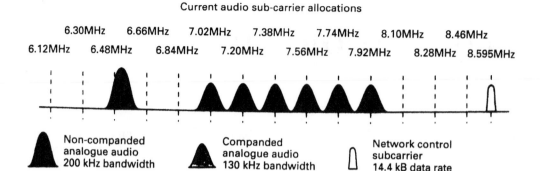

Current audio sub-carrier allocations

| Non-companded analogue audio 200 kHz bandwidth | Companded analogue audio 130 kHz bandwidth | Network control subcarrier 14.4 kB data rate |

Figure 9.2   *Astra satellite analogue audio carriers (Astra)*

primary audio carrier above which there will be several more or at least the potential for more. The Astra system gives a good account of this and Fig. 9.2 illustrates the frequencies and bandwidths involved.

This then gives rise to the need for a tuning system for audio as well as the normal channel (vision carrier) tuning, i.e. having selected a channel, which audio do we wish to listen to? Furthermore, these carriers can be listened to singly and simply split to both channels of a stereo receiver, for example, where mono sound is broadcast such as Eurosport on Astra and Eutelsat and each carrier is used for a simultaneous selection of commentary languages. Alternatively carriers may be paired to provide stereo audio (and stereo carriers can provide Dolby Surround and Pro-Logic data as well), e.g. 7.02 MHz and 7.20 MHz are respectively L and R audio for the TV channel on a typical Astra channel.

What are all the other carriers used for then? Predominantly as radio stations, either mono or as pairs of carriers to provide stereo radio. It is a fact that many satellite TV users are largely unaware of the presence of radio stations effectively for free on their systems and use the facility rarely. For those who live in a poor reception area for terrestrial f.m. radio, satellite radio provides the answer. In the UK, BBC national networks are available from the Astra satellite.

Certain service providers have chosen to use the subcarriers to carry digital services. This can lead to an increase in the number of available services

as can be seen in Fig. 9.3 a and b. Other services use the space to transmit digital data for a variety of purposes.

Figure 9.3a   *Astra satellite digital (ADR) logo (Astra)*

Digital audio via satellite

As can be seen from Chapter 14 (MPEG), the use of digital signals is becoming more and more widespread. If you simply convert an analogue signal – here audio to digital – to represent it accurately requires an enormous amount of data (so much that to transmit it would be impractical). Therefore data compression or reduction techniques are used. In the case of the ADR (Astra Digital Radio) system highlighted in Fig. 9.3, the reduction system used is called MUSICAM

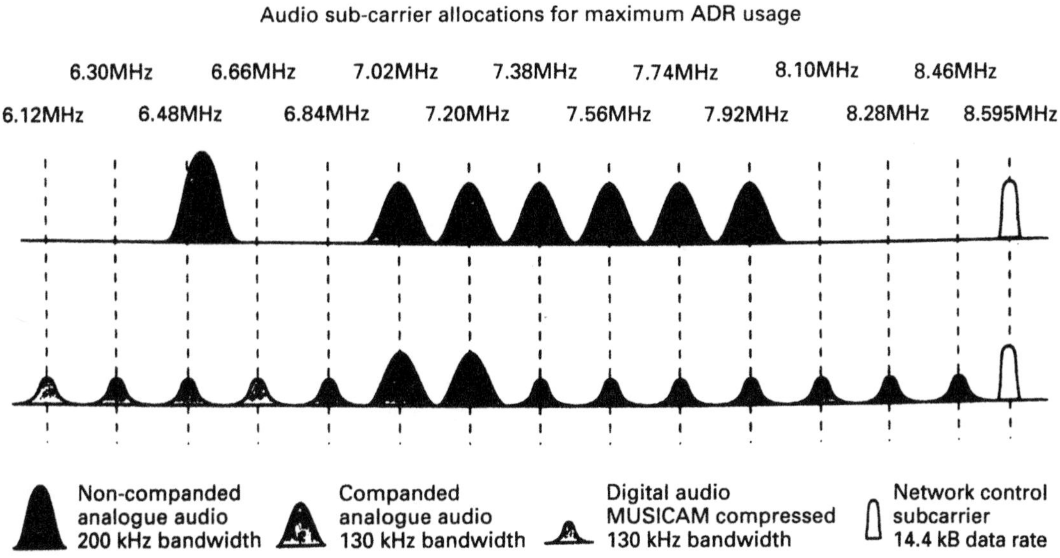

Audio sub-carrier allocations for maximum ADR usage

6.30MHz	6.66MHz	7.02MHz	7.38MHz	7.74MHz	8.10MHz	8.46MHz

6.12MHz	6.48MHz	6.84MHz	7.20MHz	7.56MHz	7.92MHz	8.28MHz	8.595MHz

	Non-companded analogue audio 200 kHz bandwidth		Companded analogue audio 130 kHz bandwidth		Digital audio MUSICAM compressed 130 kHz bandwidth		Network control subcarrier 14.4 kB data rate

Figure 9.3b   *Astra satellite digital (ADR) audio carriers (ASTRA) (continued from previous page)*

(Masking pattern adapted Universal Sub-band Integrated Coding And Multiplexing). This is also widely used in other audio fields. The reduction process uses mathematical algorithms, which are discussed in Chapter 14. The options that the programmer is provided with are large. With four Astra satellites, 64 transponders, there is the possibility of 768 carriers giving digital stereo pairs plus 64 companded pairs for television. However, what this means is that we need completely separate audio processing circuitry to handle the signals: decode, decompress and D–A convert. Currently, such services provide their own hardware which are also linked to encryption hardware to collect subscriptions. This means that little servicing experience is currently available, but the procedures will be akin to those used in mini-disc and DCC (see *Servicing Audio and Hi-fi Equipment* by the same author) and these are outlined in the chapter on MPEG.

One should be very careful not to confuse the processing of signals broadcast digitally with that of signals broadcast as analogue but processed digitally in receivers as discussed below.

Typical symptoms

Figure 9.4 provides a fault finding tree for satellite audio problems. Hopefully it will direct you to areas of the book appropriate to your symptoms. One point to bear in mind about satellite audio circuits is that it can be easy to miss a fault. Even if you haven't had a problem reported, it is wise to check all carrier frequencies as it is perfectly feasible for one to be missing whilst all others are OK. As an example, 7.02 MHz missing in the UK on Eurosport on Astra would mean no English commentary but this is unlikely to bother many Germans or Dutch – although they may notice no L audio on many of their stereo channels! Alternatively an unnoticed missing 7.56 MHz carrier may not show any problems on TV channels' audio but happen to be the mono carrier for the user's favourite radio station.

Watch out!

As there are complications to satellite receiver audio operation, it means that users are often the

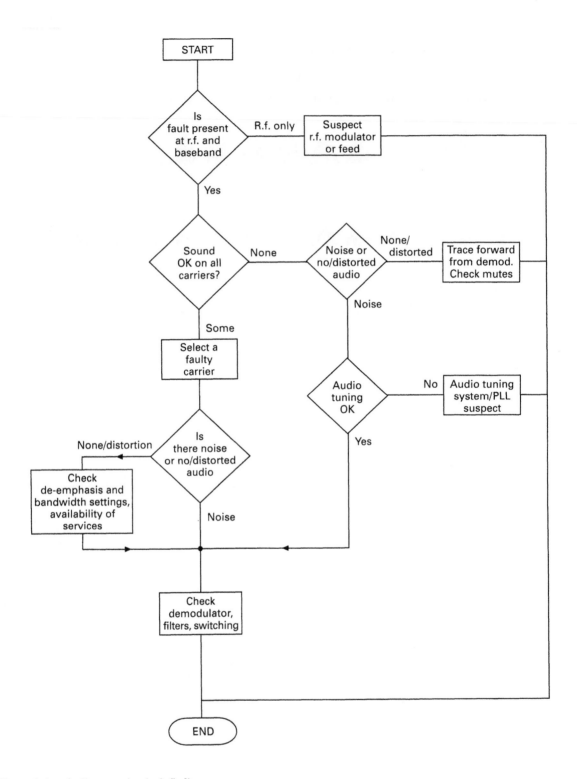

Figure 9.4 *Audio processing fault finding tree*

cause of 'faults'. Incorrectly set audio carriers can mean the wrong sound for the picture and the most common occurrences of this are again with Eurosport. The usual complaint is something along the lines of 'English voices from the left hand speaker and German from the right'! The simple answer is that the channel has been preset to use stereo audio carriers instead of the required language's mono one.

Many receivers now incorporate their own volume controls which, if varied, will result in no sound on any satellite channel or indeed a different level between satellite and terrestrial levels. The plethora of facilities available on any equipment incorporating Dolby Surround or Pro-Logic* usually gives users much to get wrong – these facilities are now available on many satellite receivers in the mainstream market and some common faults are discussed near the end of this chapter.

The audio path

Consider the circuit in Fig. 9.5 – a first generation IRD which saw massive market penetration being badged by many brands. We can see some rather novel points in the i.f. stage.

Out of the tuner comes demodulated video (as mentioned in Chapter 8) but the audio is still at i.f., i.e. undemodulated. Naturally therefore we need filtering circuitry to separate the two effectively. In the figure this is C87. This passes s.i.f. but blocks the video signal – the video signal passes via R116 to the video signal paths.

The i.f. signal is fed into pin 1 of U18 which is the superhet circuit for down-converting the tuned audio to a pair of second i.f.s of 10.52 MHz and 10.7 MHz (typical frequencies). It is not unknown for both left and right i.f.s to have the same frequency in older, non-PANDA receivers. 10.7 MHz is noteworthy for being the same i.f. as used in f.m. radio.

Audio tuning

We can now see that we have a PLL tuning system (U17 in Fig. 9.5) and demodulator (U1) to obtain and detect the desired audio(s).

This part of the circuitry is relatively straightforward to check and fault find. Where you have problems with no audio (probably just noise) and the problem appears to be this early in the sequence, then one of the first places to check is the audio tuning voltage. This appears in Fig. 9.5 on the top of D19A, a varactor diode and the d.c. here should vary as you vary the audio tuning. You will probably find that it is stuck either at maximum or minimum due to a problem with the PLL tuning circuit, here U17. The mixer IC itself is likely to be very reliable. There are many possibilities for failures with the PLL area. Chapter 7 discussed PLL tuning systems and all the points made there apply here.

The left and right channel i.f. outputs from the mixer are thus bandpassed to the demodulator. In Fig. 9.5 the ceramic filters used are clearly visible (X3–6).

In any i.f. system, it can be very difficult to determine whether your signal is present or not. An oscilloscope on the inputs to the demodulator (pins 11 and 14 of U1 in Fig. 9.5) will show very low level, noisy signals. You should just be able to discern some movement in them. The best test is probably to 'scope all the way from the mixer outputs checking through the (ceramic) filters and comparing results. Remember you effectively have two identical stages and so a fault in one means that you can compare with the other.

Switching audio bandwidths

On many receiver designs, it is possible for the user to switch audio bandwidths. Basic, Astra only designs often do not have this facility as such – just preset for PANDA on all carriers except primary audio. However, for use on other satellites' transmissions it is necessary to be able to select the various bandwidths used – typically 130 KHz, 280 KHz and 900 KHz. Should the wrong bandwidth be selected either by incorrect use or a fault, then the symptoms will be either very low sound with excessive noise where too wide a bandwidth is selected or distorted, overloaded sound with poor response if the i.f. bandwidth selected is too narrow. There is an obvious interaction with the correct de-emphasis being selected. Checking the

---

*Dolby, Dolby Surround and Dolby Pre-Logic are all trade marks of Dolby Laboratories.

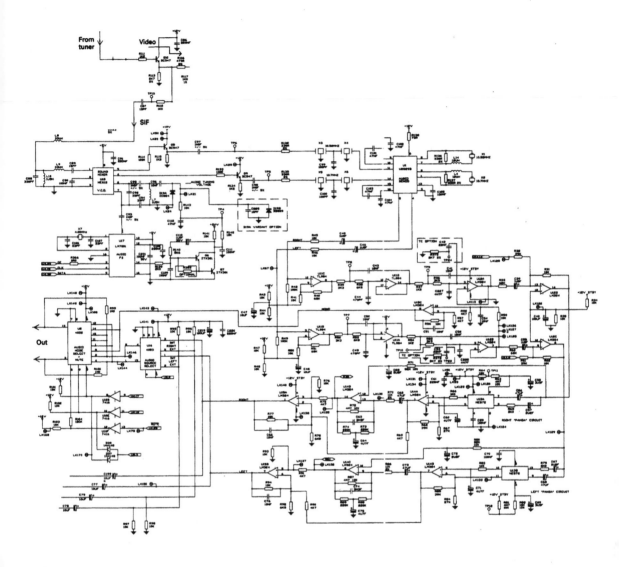

Figure 9.5 *Audio circuit from first generation IRD (Pace). This circuit is used extensively in the text to help explain signal path and fault tracing*

broadcasters' channel lists will again tell you the correct bandwidths and de-emphasis specifications to be used.

The bandwidth switching will be done in one of two ways dependant on receiver age. In earlier, analogue designs, the obvious changing of passive component values, traditionally in i.f. amplifier feedback circuits, is the order of the day. In later designs incorporating digital domain signal processing, it's all done 'in digits' within the IC.

Demodulation

It is far more likely that you will have a problem here than in the mixer stage preceeding it. Clearly

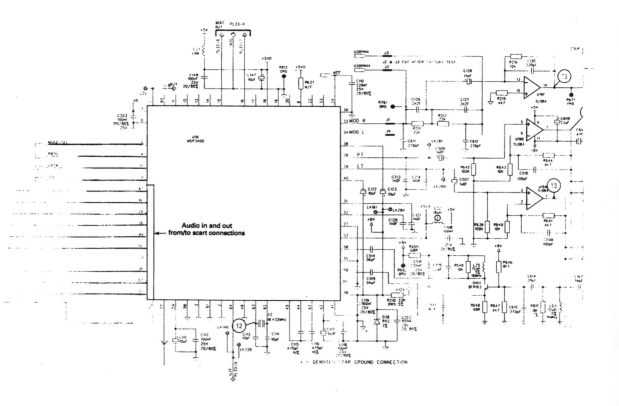

Figure 9.6 *Current analogue radio processing circuit – an IC that processes in the digital domain, the MSP3400 (Pace)*

the test that will have brought you to this point is a 'scope check on the demodulator output – pins 4 and 5 of U1 in Fig. 9.5. If one or both is missing, your problem is here or before. Aside from a total IC failure (or loss of supply) it is unlikely that you would lose all audio. One channel or certain carriers are the likely causes of being missing or distorted. By utilising the various audio routing modes of the receiver you can help to pin down where your problem might lie. Listening to a suspect carrier in mono mode might show that it is apparently fine, thus proving the tuning to be OK, but in stereo mode it will not appear. You need to combine these results with the design of the circuit to determine what it tells you!

Ceramic filters, here X1 and X2, are generally very reliable although can give rise to noises. The

most likely cause of any demodulator borne failure is the IC itself. Where we have far more integrated designs such as that in Fig. 9.6, some of these considerations become of little significance.

After demodulation

Whatever audio has been selected, mono or stereo, we have to provide it to two channels to process (unless the receiver is mono of course!). The following stages will clean up and amplify the signals but will also provide the appropriate de-emphasis and where appropriate, companding processing. We are now considering from C45/6 forward in Fig. 9.5.

De-emphasis

We need to appreciate that during processing, frequency modulation and transmission, the signal will be degraded. This will not be a linear effect – certain frequencies (higher ones) will be far more affected (by noise) than other (lower) ones. Were this to go unchecked the end result would be rather poor indeed. With satellite TV transmission, we have seen that vision is also frequency modulated and thus would suffer similarly and needs similar treatment. The specifics of video de-emphasis are covered in Chapter 8. We can largely overcome these losses by, during the transmission process, artificially boosting the affected, higher frequencies taking them well above the noise floor and then reducing them during reception.

The boosting during modulation is called pre-emphasis. The counter effect during reception is called de-emphasis. As you can now see, this takes place following de-modulation.

The circuitry in question from Fig. 9.5 is based around components U15C for the 50US mono audio (primary audio carrier).

The de-emphasis system or factor used during reception needs to be the same as that used during transmission. There are common standards and many receivers will have the facility to select more than one. Mono signals will usually use different emphasis than stereo ones. If a receiver is designed primarily for one market (e.g. Astra) it may well be restricted in this ability, as indeed is the one in Fig. 9.5. (Circuitry around U12, 13 and 14 with de-emphasis in U15a and b.) Astra use a *companding* system called PANDA (a trademark of Wegener Communications).

Companding

This is a composite word made from 'compressing' and 'expanding' which gives away its modus operandi. As we saw from Fig. 9.2 the bandwidth of the carrier at 6.5 MHz on Astra satellites is greater than the others. It is the only one that is not companded. Companding allows wider dynamic range for a given bandwidth.

The PANDA circuit consists of an expander containing a variable gain cell and a rectifier. The output of the rectifier controls the gain of the cell such that it is proportional to the average level of the input (U13). The output of the cell is converted from current to voltage and low pass filtered (U14).

Servicing

In terms of reliability, these areas tend not to give too much trouble. However, where a fair number of op amps are involved as in our working example circuit, loss of signals is reasonably common, usually due simply to open circuit ICs or paths. However, we should not overlook the possibilities of incorrect or missing biasing or feedback. Take as an example the circuitry around pins 8 and 9 of IC15. Loss of feedback due to R56 going open circuit or the output pin 8 being dry jointed will kill all signal through the device.

In theory, all manner of nasty response type problems could occur in these circuits, but in practice they don't. The guidance on the process and operation of the circuits is thus more valid than speculating on possible failures.

The modern approach to audio processing

Like many things in modern electronics, we've gone digital! It is quite reasonable in a modern receiver such as the audio circuit in Fig. 9.6 to have a large digital signal processing IC which takes an audio i.f. input and provides the required stereo analogue outputs. Figure 9.7 gives an overview of the structure of the processing within the IC. In such circuits, which are extremely integrated, your options for details fault finding are somewhat limited. However, it is in such circumstances that failing to make the prudent checks and thus condemning the IC due to its large size and all-encompassing nature, can lead to costly and annoying mistakes.

Check for supplies – here on pins 7, 18 and 39, clocks/oscillators such as on pins 62 and 63, resets and grounds. The device is controlled via the data bus at pins 9 and 10, and so activity here should be confirmed. More detail on fault finding buses is included in Chapter 11. Check for the presence of audio i.f. at pin 58. If all appears to be OK and d.c. voltage checks fail to provide conclusive proof of the failure, then you are at liberty to suspect the IC itself. It is not unknown for the device shown to fail with a variety of

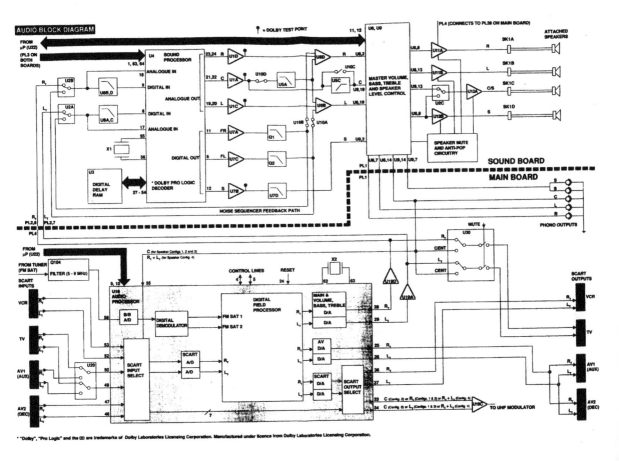

Figure 9.7 *MSP3400 IC block diagram. Also shown is the block diagram of the Dolby Surround and Pro-Logic circuitry (Pace)*

symptoms but usually on the lines of no audio at all or just noise.

Audio switching

As with the vision signal, we need to route audio signals. This can be quite simple as in Fig. 9.5 (where we simply switch between mono or stereo and internal or external audio) but expands when we consider receivers containing sound processing such as Pro-Logic and where other a/v sources (e.g. hi-fi VCR) are routed through them. In other words, routing has to take place even when the unit is in stand-by.

One of the areas where audio switching is required, as with vision, is with feeding through a decoder, although audio decoders, particularly in the UK market, are much less common than for vision signals. The principles are the same as for vision, the take-off point being the critical point.

The possibilities for problems are great here. Again it's just basic signal tracing in and out but also checking the switching logic – bear in mind the direct connection to system control.

Muting

The degree to which muting is implemented on any given receiver will vary. Some older or cruder designs may not have it at all. The uses will be to prevent noise when there is no audio present or when tuning and indeed at switch-on and off to prevent 'plopping' through the speakers. It is often overlooked when tracing 'no audio' faults. The switching will come from system control but if it is erroneously being activated, the problem is more likely to be the method of detection used to switch the mute on that is in trouble. The most obvious method is detection of sync pulses on the incoming video signal: direct attention here if the wrong logic level is emanating from system control. Muting will invariably be solid state in satellite receivers and so the devices themselves will likely be very reliable.

Muting can be seen in Fig. 9.5 at U2.

Dolby Surround and Pro-Logic

The best defence against reported problems in these areas is a good understanding of what's involved in the systems. It is highly unusual to get true faults here! Most complaints centre around loss of audio from some of the speakers. In turn, this is usually due to having pressed the wrong button or by using unsuitable software. In the case of satellite, the software is usually off-air programming but equipped receivers allow routing of VCRs, etc., through them to facilitate use of the decoder and audio amplifier. It has first to be established, then, that the material viewed when the problem occurred was either Pro-Logic or Surround encoded. Most programmes will be labelled at their start to indicate this. All encoded material must bear the labelling shown in Fig. 9.8. A very large proportion of BSkyB material is Pro-Logic encoded (all such material can be played via normal stereo circuits without problems) but in the event of being unable to confirm this, you could play a video tape through an external input to test the unit.

Dolby Surround

This is a passive surround system but with information encoded during the recording process along with Dolby noise reduction (any stereo signal can be Dolby Surround or Pro-Logic encoded). Figure 9.9 shows the block diagram of a surround decoder. A decoding matrix is used to recover the single surround channel which is fed as a single channel to two rear speakers.

Dolby Pro-Logic

Two extra channels are achieved in this active surround system. A centre channel is used to concentrate the viewer's attention and so the single centre speaker is placed near the TV screen. Surround is again applied via a rear channel. Pro-Logic gives a much greater effect and allows much wider tolerance in positioning of speakers than the passive system.

Problems

There will be user adjustments allowing the time delay between front and rear speakers to be adjusted; there will be switchable modes to force Surround or Pro-Logic (thus causing problems with no sound from certain speakers if material is not suitably encoded); there will be a phantom mode to allow use without a centre speaker; there will be facilities to use the TV speaker as centre channel. Taking the latter as an example, if you inadvertently switch to this mode when you have no external speakers connected, you will get very odd sound on the TV at best and at worst none at all! See Figure 9.14.

Audio power amplifiers

The inclusion of such circuits, to drive loudspeakers directly is primarily intended for designs with the audio surround processing facilities such as the Pace MSS1000 as in Fig. 9.11. The designs tend to be simple, IC based ones (Fig. 9.12) and servicing is therefore straightforward. It is not my intention to cover all servicing aspects of audio amps here, detailed reference can be made to *Servicing Audio and Hi-fi Equipment* by the same author, but some prudent tips follow.

Failures are usually more than a no-sound symptom (Fig. 9.10). The IC (U11 and U12) contains all the power handling capacity of the amp. It receives a supply (or supplies), here on pin 12 and audio on pins 3 and 6. Outputs are on pins 11 and 13. If there had been a fatal failure, the IC having gone short circuit, it would load the power

All passive Dolby Surround decoders (or equipment incorporating these decoders) are marked with the logo:

Active Dolby Surround decoders and equipment incorporating Pro Logic circuitry are marked with the logo:

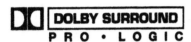

Figure 9.8   *Dolby Surround and Pro-Logic logos (Dolby Labs)*

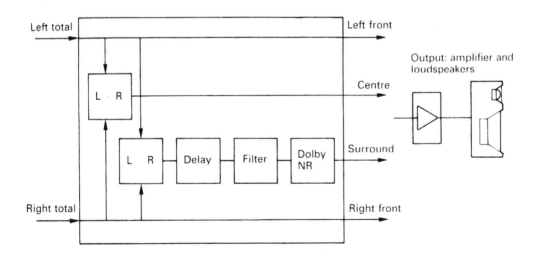

Figure 9.9   *Block diagram of a passive Dolby Surround decoder (Bang & Olufsen)*

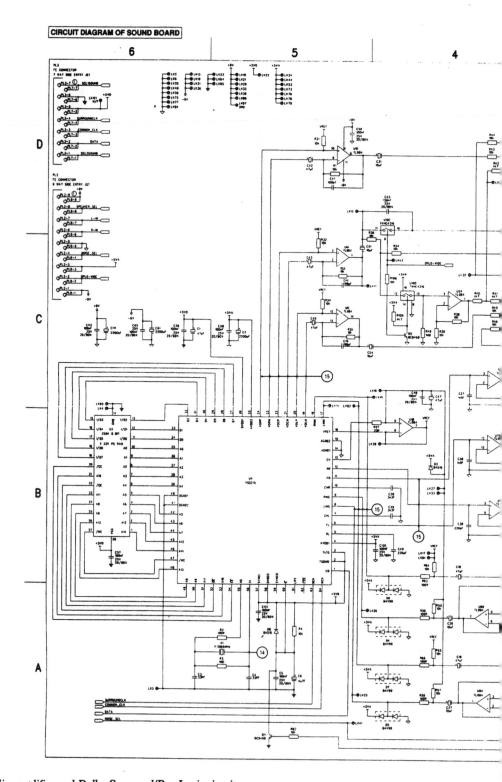

Figure 9.10 *Audio amplifier and Dolby Surround/Pro-Logic circuitry.*
Note the option to include a cooling fan for the audio amps (Pace)

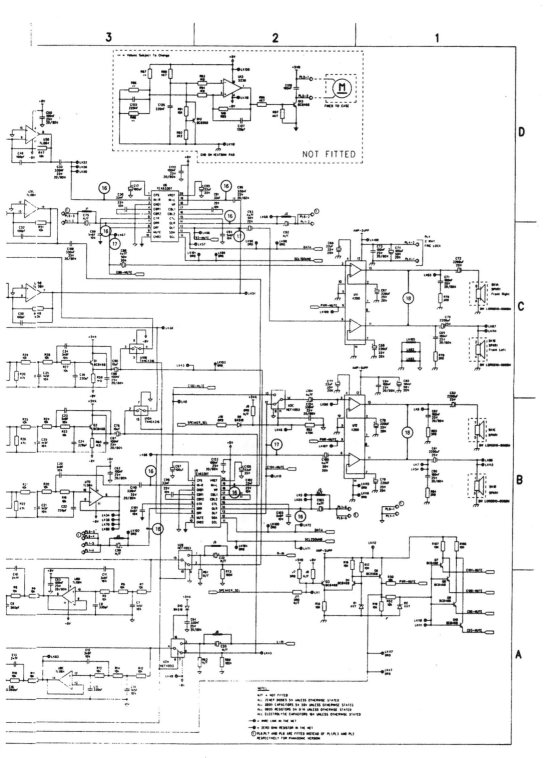

THIS CIRCUIT HAS BEEN SPLIT TO FACILITATE PRINTING

Figure 9.11   *Modern receiver incorporating Pro-Logic circuitry and audio amplifiers*

Figure 9.12   *Photo of audio process and output PCB from receiver shown in Fig. 9.11*

supply intolerably and so the latter would enter protection mode, in this case pump. Distortion can be easily checked by monitoring input and output signals – inject an external tone if necessary, but remember to do so via an isolating capacitor.

If output IC(s) have failed, ensure that all connected speakers and wiring are checked for short circuits. Reconnecting in this condition will lead to immediate re-failure. Also check any fan cooling circuitry to ensure that it is working – the failure may have been due to overheating.

Aside from the ICs, the other likely failure in this circuit would be the coupling capacitors C70/2/81/3. Low audio would be the most likely complaint.

Interfacing with the outside world

Naturally, the way in which the user obtains audio from the receiver will vary as it will with the video signal. We discuss the r.f. modulator in Chapter 12 and will not repeat ourselves here, but suffice it to say that a feed from the audio switching will be fed to the r.f. modulator to provide sound with the vision to be fed to the TV via its aerial socket.

Useful fault finding information can be gathered (as it can with vision faults) by checking the sound at r.f. and baseband (via the a/v out connections – usually scart) and seeing if the symptom is present on both. For example the audio to the r.f. modulator may be buffered via a transistor to the r.f. modulator whereas the scart to the TV is direct from the same origin. Therefore is you have sound at scart but not r.f. you can immediately rule out problems prior to the buffer and will likely find a fault with the buffer or r.f. modulator (or indeed connections between them).

With stereo receivers, a pair of RCA phono line level outputs will often be provided (more on Pro-Logic units) – see Fig. 9.13 – to provide a convenient way of connecting to a hi-fi amplifier. This would be very useful if the user has a mono TV or one with poor audio and wishes to use the superior performance of a hi-fi. Similarly they may wish to use an external Dolby Surround or Pro-Logic amplifier to make use of the widely available software on satellite.

As with all connections to the outside world, scarts, phonos and r.f. sockets are prone to wear and breakage. Usually in all modern equipment, and satellite receivers being no exception, such connectors tend to be soldered and mounted directly to the board – those units with a higher

Figure 9.13   *The audio interconnects available from such a receiver*

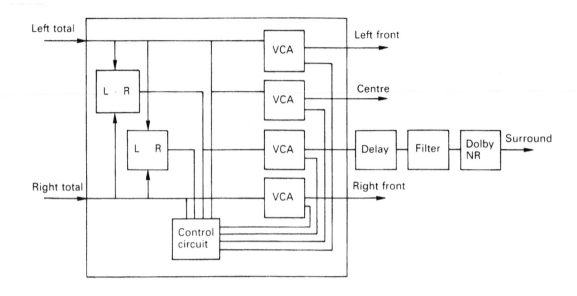

Figure 9.14 *The Pro-Logic decoder principle (Bang & Olufsen)*

build quality (and consequently higher price) will have such items hardwired. The inevitable consequence of the cheaper alternative is the occurrence of dry joints and broken print on the connectors. This can be the answer for many intermittent faults.

Audio feeds will also be present to any decoder connections provided on the receiver. Many readers will be familiar with connecting vision decoders such as video crypt but not many audio scrambling systems are in widespread use. However, the ADR system described earlier in this chapter may well require the use of baseband audio feeds to decoders, although the preference is for separate receivers with their own tuners.

10

POWER SUPPLY CIRCUITS

The power supply is an area of any piece of equipment with which an engineer should be particularly familiar, because it is generally one of the most unreliable sections of all. There are many reasons for this: primarily high voltages and currents giving rise to lots of heat, which increases the rate of failure of components and increases the likelihood of dry joints. Breaks in printed circuits occur because of the heavy components used and the high temperatures and currents present. Failure elsewhere in a unit can also result in damage to the power supply, which feeds all stages of a unit. As well as the obvious faults causing 'no-go' through missing rails, the power supply section may give rise to niggling faults apparently totally divorced from it, and caused by excessive ripple/poor regulation on a particular rail – perhaps intermittently for good measure!

With a satellite receiver, particularly mainstream ones, we have a unit consuming typically 35 W (some up to 160 W!). This is akin to that of a VCR or even small television, in a case somewhat smaller. The ventilation therefore is poorer. Add to this the irresistible temptation to stack the receiver on top of aforesaid VCR or TV and maybe also in a cabinet or with a decoder box on top, and you can begin to have sympathy with the power supply units (PSUs)!

It is very important to avoid this scenario. Educate customers as much as possible. They will often respond if they think that the unit will be more reliable and cost them less to maintain!

The PSU is the root cause of many a symptom with a satellite receiver, and the symptoms are often puzzling. The golden rule is that whenever you have a fault in a particular area, always check the level and cleanliness of a supply sooner rather than later. Take time to fully assess the symptoms – there are often more subtle symptoms beyond the obvious and these can direct you to the PSU which is often the only common link.

Before getting too deep, however, we shall start at the beginning and look at all power supply factors.

Safety

It is essential not to under-estimate or disregard safety. Working around mains power supplies represents a danger to yourself and the potential to make the unit dangerous to use in future. If you are not competent, you should never endanger yourself or others by working on mains powered equipment. Most units today use switched mode power supplies which have the added hazard of a live chassis primary circuit. When working on such units, they must be powered via an isolating transformer. All mains powered equipment must be tested for electrical safety following repair before return to the end user.

Mains connections

Many receivers have detachable 'figure of eight' mains leads, others are permanently anchored inside the appliance. Either way, attention has to be paid to them.

The mains lead may seem to some unworthy of attention in a fault finding book: nothing could be further from the truth. Mains leads are critical safety components, and great attention should be paid to them and their connections. This means that every item serviced should have the mains plug connections checked and remade as required; the fuse value checked; the fuse-holder connections checked for tension and good contact; and the overall condition of the plug confirmed to be safe and free of cracks, holes or signs of overheating. The mains lead should be checked for kinks, breaks in insulation or other damage. If any fault is found the lead should be replaced as should the mains plug itself if defective. Recent legislation in the UK requires the fitting of mains plugs with sleeved pins on all goods sold – a point easily overlooked when refurbishing equipment. Damage at the ends of leads may possibly be cured by cutting out the affected section, provided it doesn't render the lead too short for the customer's use. Replacement leads should meet the manufacturer's specification as with all safety components. Another important factor when replacing leads is the need to replace all cable ties and other restraining devices such as blobs of glue, as well as all insulators. Commonly where a mains lead is connected inside a unit, either on a mains switch or at a mains transformer/input PC board, a cable-tie is attached to prevent the leads shorting out should one or both break off, and glue or some other retainer provided to hold the lead in a particular position. Neglecting to refit these safety devices invalidates the original safety standards met by the unit. Failure to replace insulating covers or boots over connections and fuses could result in the death of you or the user of the equipment.

Having checked these essential points it's good practice to recheck for damage inflicted during repair prior to returning the unit to its user. The initial check is to protect yourself from an unexpected shock from an incorrectly wired plug connection or other such fault. It is amazing to see how some mains plugs are wired by users, even in the presence of a wiring diagram in or on the plug! Re-checking at the end of the repair shows up any damage to the mains lead caused by its being trapped, abraded or stretched during service. Breaks in the lead conductors usually occur where they enter the plug or unit or at 'kinks' where the lead is under most stress, or where someone has tripped over the lead, stretching it. They can be repaired by cutting out the damaged section, subject to the lead being of sufficient length afterwards, as stated above.

Mains filtering

Once inside an appliance the mains feed will be treated in different ways depending on the design of the unit in question. Because different countries have different standards for mains filtering, mains power may be applied straight across the transformer, possibly via a mains switch; or may be fed through filter chokes or other devices if standards require – or if the unit produces noise that would upset the mains supply. These components represent another possible failure in the power supply chain, so if no voltage is present at the transformer or switch an O/C, broken or dry-jointed choke could be responsible.

A switch suppression capacitor, typically of 0.1 μF may be used, and can go short circuit with explosive results. It should be replaced rather than just removed from the circuit.

This component can be the cause of random mains fuse failure – if the fuse has gone open but is not visibly black. On many occasions there has been a modification to lower the value of the capacitor across the poles to prevent such random problems. This must only be done as an approved modification from the manufacturer.

With switched mode PSUs, there is a need for more complex filtering here to prevent noise from the power supply in the receiver being fed back to the mains supply system. At its simplest, this is a bifilar choke. Little in the way of faults are experienced here.

Fuse types

It is important to replace fusible components with the correct value and type. Quick-blow types are usually identified by their single strand of wire in the glass envelope. Semi-delay types are least common, and anti-surge fuses, denoted by a 'T' following the value (e.g. 3.15 AT), have a spring at one end of the wire.

Table 10.1 *Colour codes for thermal fuses*

Brown	72°C
Cream	84°C
White	91°C
Red	100°C
Blue	109°C
Cream	121°C
Brown	128°C
Red	141°C
Grey	152°C
Lime	167°C
White	184°C

Whickman fuses, or ICPs (discussed later) are marked with a code which equates to 1/40th of their rated current in milliamps. Therefore ICP-F10 is a quick blow 400 mA device. The other type is the N type, e.g. ICP-N15 = 600 mA. Fig. 10.1 illustrates these devices.

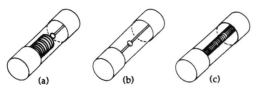

(a) Dual-element delay fuse construction. Easily recognised by the helical tension spring

(b) 'Solder-blob' type delay fuse

(c) Spiral element delay fuse

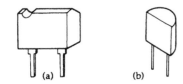

ICP packages. (a) ICP-F type. (b) ICP-N type.

Figure 10.1   *Glass fuse and ICP designs and constructions. Great care should be taken to ensure that the correct type of device is used for replacement purposes*

Thermal fuses are quite common; their uses are discussed throughout the chapter. Table 10.1 gives the colour-codes for these devices, in which some colours can mean two different things! If the manufacturer's data does not specify the temperature, the application gives a good indication as does any numeric coding on the body of the device.

Linear PSUs

In simple PSUs the mains voltage is stepped down, rectified and smoothed, then (usually) applied to a regulator or stabiliser of some sort. In older equipment this takes the form of a transistor series-regulator circuit like that shown in Fig. 10.2. More recent circuits use IC regulators, ranging

from simple three-leg, one-output devices to multi-pin, multi-output variants. All these types will be examined in this chapter from the troubleshooting point of view.

Primary circuit

Following what is commonly termed the mains input circuit (the parts of the power supply covered up to this point) the primary power circuit is met.

Here mains voltage is applied to the primary winding of the mains transformer. Power on/off switching is not necessarily carried out in the primary circuit; in modern designs where a 'standby' mode is adopted, rather than a complete 'power down', switching is done on the secondary side of the transformer.

Power switching

With satellite receivers, it is usually the case that there is no physical mains isolation switch. Those designs from more specialist (non-Astra) manufacturers may well have and will doubtless be switching mains at the primary of the power supply circuit. Most mainstream designs though simply have a stand-by mode and not only have no mains switch but no front panel controls either – all commands are from the remote handset! It has to be stated that any mains switch is a critical safety component and therefore must be replaced if determined to be faulty in any way – latching poor or soggy, arcing, intermittently open circuit, etc. It must only be replaced with one approved by the manufacturer of the receiver.

Primary fusing

As the mains enters the unit the supply is protected by a fuse in the mains plug (or distribution board if the plug itself is unfused) which will, of course, have been checked to be of the correct rating by the engineer before service or installation. This offers protection in the event of, for example, a cut mains lead and for satellite equipment is typically 3 A (UK), but each individual unit should be fused to the manufacturers' specification. A primary fuse is usually included in one form or another which will be of a lower rating than the fuse at the plug or distribution board. This is typically of the glass anti-surge type (see Fig. 10.1) to prevent random failure in the event of small power variations on the mains supply system, and failure under the heavier switch-on surge currents. Another common form of protection takes the form of a thermal fuse at the primary winding of the mains transformer. This may or may not be shown on the manual's circuit diagram, so look at the transformer to check for its presence. If there are three primary connections to the transformer it is likely that one is neutral, one live and the third the junction point of the thermal fuse. Typically these are arranged in a 2 + 1 format, and there should be electrical continuity through the 'pair', via the fuse. Some thermal fuses are more useful than others. For example, if the fuse is removable – generally being embedded within the transformer – and replacements available, the job is simple so long as the cause of failure is established. Some fuses are not replaceable, necessitating the replacement of the whole transformer, others don't seem to be very effective as it has been known for the transformer primary to go open circuit while the fuse remains intact!

Thermal fuses may also be fitted external to the transformer, typically across the two terminals mentioned above. External thermal fuses are commonly small metal-canned torpedo-shaped devices with a colour code (Table 10.1) to indicate their operating temperature.

Mains transformers

Transformers are generally reliable components. It is usually easy to identify a faulty transformer and to prove the diagnosis: few problems are encountered in practice. Apart from failure of the thermal fuses described above the only other 'common' fault afflicting mains transformers is that of O/C windings. With an O/C primary winding the symptom is likely to no operation of the unit at all. If the transformer has but one secondary winding, and *that* goes O/C, the same applies.

Occasionally a transformer develops short circuit turns, usually resulting in the opening of a fuse. This may not happen until the insulation of the transformer's windings breaks down in turn, possibly due to the overheating effect of an over-

load. In a recent case in the author's experience a mains transformer was damaged by a short-circuit in the audio output IC of a hi-fi system which was in turn damaged by a short circuit in the speaker wiring in the customer's house. Such a fault can cause further heat to be developed and thus lead to a fire risk: much depends on circuit design and transformer construction. If a secondary winding develops shorted turns the output voltage can change, causing all manner of strange symptoms and possibly damaging components. Interwinding shorts can also have disastrous results, but in most cases the fusible device will operate before too much damage occurs.

If in any case the output from a secondary winding increases it must be assumed that some damage has been done to the secondary circuit: check as appropriate. Unless the increase has been great it is unlikely that the rectifier(s) will have been affected, but with the rectified supply voltage having risen it's wise to check out regulator circuit components such as reservoir capacitors, transistor/IC regulators and Zener diodes.

Mechanical noise from any wound component carrying high voltage, currents or frequencies is a common problem. The treatment for an excessively noisy component has to be carefully considered when it is in a position critical to safety: this mainly applies to mains transformers. It is possible to suppress some noises by applying a plastic sealant or wood glue, but care must be taken not to allow these near any area where heat is developed as a fire risk could result. If the noise is excessive and resists treatment, replacement of the transformer is the only safe course of action.

We shall consider specifics of switched mode power supply transformers later in the chapter.

Secondary circuits

We shall now look at the various component parts of secondary circuits. This is with an eye to linear power supplies but many switched modes use identical circuits in their secondaries and so by reconsidering ripple frequencies, all information applies equally to them.

Secondary fusing

The presence and type of secondary fusing depends greatly on circuit design. Very often

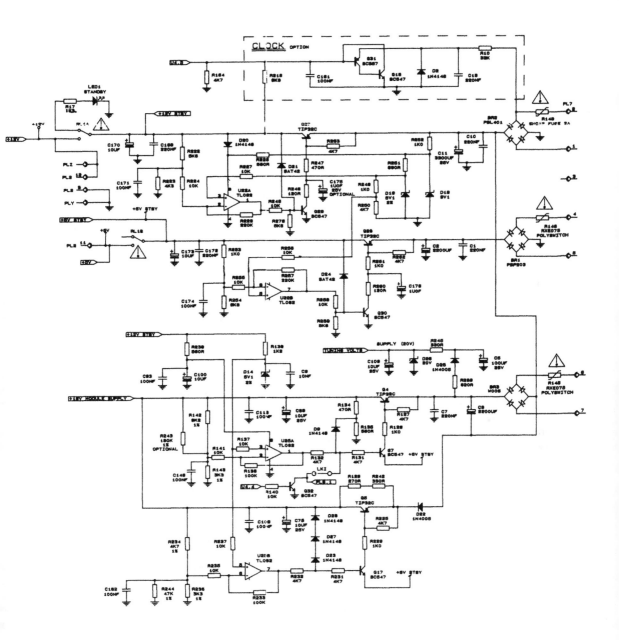

Figure 10.2   *This linear power supply illustrates many examples from the text (Pace)*

a fusible device is fitted between the transformer's secondary winding and the rectifier. Other circuits may have the regulated output fused, leaving the primary fuse to protect against any fault in the components between the transformer and the secondary fusible component. In odd cases with this type of circuit (or where an incorrect value fuse has been fitted) damage to the transformer can result, so checks here should not be overlooked if, after tracing the fault in the secondary, there is still no supply from the transformer.

Many devices can be found in the protection role. The obvious fuse(s) are easily enough checked. Also found in more modern units are fusible resistors, typically low value devices designed to operate like a fuse. They are also usually non-flammable types and (as with any fusible component) they should be replaced with the specific type approved by the manufacturer.

Another increasingly common device in these positions is the 'semiconductor' fuse, otherwise known as a Whickman fuse or intergrated circuit protector (ICP) covered earlier.

One other form of fusible device that we encounter in satellite receivers is the self-resetting thermal fuse or 'Polyswitch'. Once opened by an excess current, it will reset if allowed to cool and so is useful for LNB supply protection. The idea has been largely supplanted by more advanced electronic methods with switched mode supplies, but with linears, this device is quite common. It is a fact, however, that each time the device operates, its threshold for operation seems to lower and so makes it more likely to operate.

Rectifiers

Depending on the required regulation of the d.c. supply and the subsequent regulator circuit, the type of rectifier circuit used varies. Readers are probably familiar with the principles of full-wave and half-wave circuits and it is not the purpose of this book to explain them again. Therefore we will consider fault conditions in rectifier circuits.

A short circuit in a secondary rectifier causes some protection device to operate. If the secondary protection is not upstream of the rectifier, primary protection should operate: this is often also the result of a rectifier becoming leaky, although the symptoms may have been preceded by a hum on the unit's output. Hum on supplies

to more complex circuits can be rather more nasty, perhaps destroying semiconductors.

An O/C rectifier causes loss of the supply if it is a single, half-wave rectifier. In a bridge formation one rectifier O/C is likely to cause the hum mentioned above and – depending on the protection arrangements – may or may not blow a fuse. Rectifiers may only appear O/C under load, reading OK on a meter.

Smoothing/reservoir capacitors

Electrolytic capacitors used in these applications are chemical devices and so have a finite life. This life is reduced by the high temperatures and currents which exist in power supplies. The capacitors tend to dry out thereby reducing their effectiveness to a greater or lesser degree. The result is a reduction of the d.c. voltage and an increase in line ripple and hum. We shall consider this in more detail later.

Split supplies

A split supply is where a pair of supplies equidistant in potential either side of ground are produced, e.g. +12 V and –12 V. These can be achieved quite simply by using rectifiers in opposing configurations in either half of the supply.

It is a result of this type of arrangement (or where such regulation is incorporated within an IC) that when one side of the supply is lost or upset, then the other will be also – a point that needs to be remembered.

Transistor regulators

These basic circuits invariably use a large power transistor in conjunction with a smaller transistor driver circuit to provide regulation of the rectified and smoothed supply from the mains transformer's secondary. The drive circuit samples the regulated output (usually from the regulator transistor's emitter) and compares it with a stable reference, typically from a Zener diode. From these inputs it produces an error voltage output to drive the regulator, thus compensating for input voltage and load fluctuations. As can be seen from

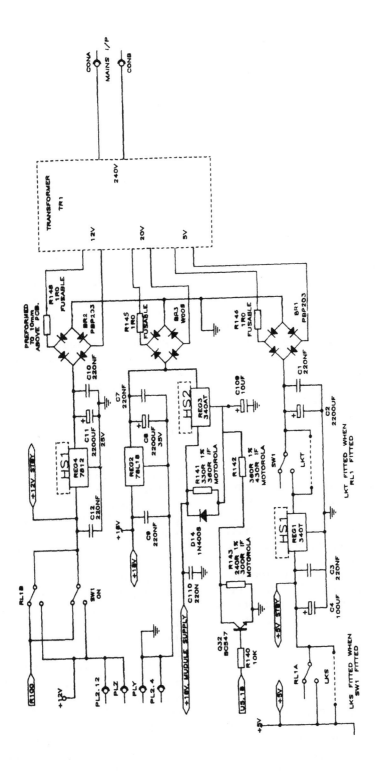

Figure 10.3 *This older, more basic linear power supply demonstrates the greater simplicity of using IC regulators (Pace)*

Fig. 10.2, the inclusion of a comparator amplifier in the series regulator circuit is common practice in satellite receiver linear power supplies. Here we use the in phase input of the comparator to apply our reference supply and the negative, anti-phase input takes a pot divided sample of the regulator output thus producing an error to drive the amplified Zener series regulator.

The faults that occur usually involve poor (or lack of) regulation or no output.

Poor regulation

This is manifest as 50 or 100 Hz ripple on the output which is also usually low until its load is completely removed, causing many problems, especially hum. This type of circuit is seldom used in equipment incorporating microprocessors, avoiding many potential problems. In transistor regulator circuits the main reasons for poor regulation are commonly split between a faulty smoothing electrolytic and the regulator transistor itself or its drive semiconductor. The former can be suspected if the d.c. input level to the transistor's collector is low. A 'scope check, or bridging the capacitor with an equivalent type, are valid checks for confirmation purposes.

No regulation

If the 'stabilised' output voltage is high – perhaps the same as the regulator's input – it could well be due to a S/C or leaky regulator transistor; in some circuits this will blow the fuse. If the problem is an inability to set the correct output the likelihood is that the main transistor's base drive from the error detector is wrong due to a fault in the drive transistor or Zener diode. Once these have been ruled out suspect the higher value resistors for O/C or high resistance: these cause a shift in d.c. conditions.

No output

Complete loss of regulated output implicates most of the same components as for no regulation. Sometimes the regulator transistor goes O/C, transferring the current burden to any parallel resistor which may be present. As a result the resistor overheats, pointing the finger of suspicion at the transistor. Short-circuits in either transistor or Zener diode may well blow a fuse to draw the engineer's attention.

IC regulators

In modern power supplies the most common voltage stabiliser is a purpose-designed IC. Single-output devices come in transistor-like three-legged plastic packages, while multi-out-put types come in multi-legged 'slab' form, similar in appearance to audio output ICs. The smaller devices are easy to deal with. If you have to work on a unit without the aid of a service manual (avoid it if possible!) their type number indicates the rated output voltage. For example, MC7805 gives +5 V, LM7812 gives +12 V, and 7918 is a –18 V type. These are common examples, enabling you to check the three pins' voltages. Expect to find a ground pin, usually (but by no means always) in the centre the regulated output on the third pin; and a higher (unregulated) supply voltage at the first pin. Using the circuit in Fig. 10.3 we'll consider some faults in IC regulator circuits.

No input, no output

Loss of regulated voltage can be due to several causes. First check for the presence of the un-regulated input – if it's missing, a protection device (in fuse, ICP or resistor form) has probably opened. Random failure of these is quite common in modern equipment. Fit the correct replacement, and bear in mind that failure *could* have been caused by some intermittent excessive-current fault. In any case, check for excessive loading, perhaps in the form of a short-circuit to ground in the regulator IC, or a short-circuit or heavy load on the output of the device, i.e. in one of the circuits being fed from its stabilised output rail.

In the circuit shown in Fig. 10.3 the lack of input to *one* regulator will probably be due primarily to the failure of the fusible resistor in question: R145/6 or 8. If so, look for a short circuit to have caused their demise – including here, a possible short on the LNB or down lead, (in the case of R145). A missing supply to a regulator input could be due to some traditional power supply weaknesses such as a dry joint on the transformer or bridge rectifier, a break in the print in the same areas, or an O/C secondary winding on the mains transformer.

No output

If an input is present and correct the loading on the regulator's output line should be checked. It is possible in practice that an excessive load (not a dead short circuit) could for a time just load down the output of a regulator before causing the related fusible device to operate. Whilst investigating such a fault, then, if the unit hasn't been operated in the fault condition for too long, the protection device could operate, misleading you to assume that it was due to your work, possibly by causing a short circuit with a probe or screwdriver.

The no-output symptom would most likely be caused by an open-circuit on the ground connection or inside the regulator IC itself, possibly brought on by an ancillary fault such as an O/C reservoir capacitor on the supply side. While most regulator ICs fail for internal reasons it's worth checking for excessive input voltage and for heavy output loading. In a circuit like that of Fig. 10.3 such a fault would probably cause failure of the feed resistor where fitted; heavy loading or *unprotected* lines could well 'blow' the regulator IC rather than the primary fuse F1. No real safety risk is involved here: the equipment designer takes into account the breakdown characteristics of the components. Dealing with this type of thing often depends more on practical experience than on theoretical knowledge.

No regulation

If the IC fails to regulate, its output voltage becomes high. The most common cause of this is a short-circuit within the IC. Again it's usually for internal reasons, but check input voltage and load current as suggested above.

As a consequence of regulator failure, downstream components may be damaged by the abnormally high voltage supplied to them in the fault condition. Check them, and bear in mind that repeated failure may be caused by intermittent or spasmodic failure of the regulator IC or its ground connection.

IC multiregulators

As we've seen, some PSUs use ICs containing several supply regulators. These often incorporate power switching pins, operated by d.c. control lines from the system control circuit. If one or more output lines are 'dead', then, check the status of the command lines as well as the points given above for single-voltage ICs.

D.c.–d.c. converters

Secondary voltage supplies, typically used in fluorescent display panels, are often provided by a *d.c.–d.c. converter* in the form of a small plastic or metal module. In the event of failure, confirm that an input voltage is present before condemning the module.

Protection Zener diodes

We have seen Zener diodes being used for their ability to regulate; another common use of Zener diodes is for protection. This may take place in the power supply where if a supply line rises the Zener diode pulls it down and causes a safety device to operate. Alternatively they may be fitted on data bus where 5.1 V devices (typically) are used to ensure that the d.c. voltage doesn't rise in the event of (for example) a power surge, thus protecting expensive ICs. In either case, failure of the Zener diodes themselves will result in loss of the d.c. supply. In the case of the bus this may render a unit 'dead' as there will be no pull-up and thus no clock/data on the bus. Before replacing the device do check for a reason for the failure – excessive loading, excessive supply, etc.

Power indicators

Most units incorporate some form of power-on indication; with modern units this may also extend to stand-by indication. Older designs may use mains neons, or bulbs powered straight from a secondary supply. These bulbs often fail due to the relatively high power and heat involved. Modern designs generally use small LEDs, sometimes bi-coloured, powered from a low-voltage supply and usually switched by the system control section. Here the behaviour of the LEDs may give diagnostic clues, unlike the old 'pilot-bulb'. Indicator faults usually stem from more serious problems such as system-control faults or missing supply line voltage. In modern designs the indicator may play a 'fault-notification' role so if, for instance, the unit returns to stand-by following an electronic hiccup do not spend time looking for a fault in the wrong area!

Survey of secondary supplies

It may be useful to consider what each secondary supply is used for – this is bound to speed up fault finding.

- 5 V supplies are used everywhere where there is logic! The microprocessor, control circuits – maybe infrared amplifier for remote control reception although sometimes this uses 12 V – etc. This supply may have a switched version but there must obviously be a permanent supply to the microprocessor and infrared receiver otherwise you would never be able to get it out of standby! 5 V may also be used for electromagnetic or servo motor polariser drives.
- 12 V supplies are again multiplicitous. This time they will supply audio and video processing circuits. One 12 V supply (or 13 V) will be used for LNB polarity control when using Marconi type LNBs. 12 V may also be used for the supply to an electromagnetic polariser.
- 18 V is used for LNB supply whatever type of LNB is used. At 17 V, it represents a polarity switching supply for Marconi type LNBs.
- A 24 V supply is used for tuning – audio and video and in Pace receivers also to tune the r.f. modulator output. In some extended range receivers (for Astra 1D) the tuning voltage is 30–33 V as is standard practice in TV and VCR tuning systems. This supply may also be used for older LNB supplies.
- A 36 V, high current supply is used for positioner motor drive.

Switched mode power supplies

This type of power supply is now widely used in satellite receivers, as it has been in TVs and VCRs

for some years. It is very important therefore to understand their operation and how to fault find them. This process is subtly different with satellite receivers, as experience has proven that many failures are chain reactions whereby blanket component replacement is necessary to prevent re-failure within a very short period of time. A plethora of power supply repair kits has become available for certain ranges, some by the manufacturer, others by third party companies. In the latter case, one has to be cautious as the contents do not always meet the specification of the manufacturer. These kits, however, invariably only relate to primary circuit failures and there are plenty of secondary problems that will need attention!

Operation

It is not the province of this book to give detailed information on the operation of switched mode power supplies. We will, however, consider the rudiments. Refer to Figs. 10.4 and 10.5.

The basis of the design is that an oscillator runs at a high frequency (several kHz) and its output feeds a pulse-width modulator. The pulse width is determined by the output of a comparator, the inputs of which are a reference preset, and a feedback of the output voltage of the supply. As the voltage out drops, so the pulse width is increased. The PWM output is used to drive a switching device which switches the unregulated, rectified mains through the transformer primary.

To start the power supply a d.c. is tapped off the unregulated h.t. via a resistor network. The oscillator runs and as the current is switched through the primary a feedback is induced in the close coupled primary winding which is used to run the oscillator in normal operation.

It will be noted that one side of the mains is connected directly to the ground in the primary circuit (hot ground). When making any measurements in the primary, the meter or 'scope ground should be connected to this hot ground. To repeat my earlier warning, the unit **must** be powered via an isolating transformer. Grounding this point will lead to a big bang!

Fault finding

A completely dead unit may have three basic causes.

Not starting is the simplest. No d.c. to the oscillator should direct attention to the dropper resistor(s) which will usually be found to be open circuit. Some designs will kill the start supply as a form of overcurrent protection – look for a transistor to ground at the oscillator end of the dropper resistors.

A dead primary is usually indicated by the failure of the fuse or input protection device. Aside from the sections common to linear supplies such as bridge rectifiers we should direct attention to the switching device and any related diodes or

d.c. coupled driver devices which will almost certainly be short-circuit. If this is the case, check low-value resistors in the emiter circuit (assuming a bipolar transistor) for open circuits, and if a capacitor is used to couple the drive to the device's base suspect that it has dropped in value – check and replace as necessary. Failure to find the latter will result in re-failure. If an IC is used to control the primary and the switching device has failed violently then replace the IC. If the stage is discrete and there has been a failure suspect that the other silicon is damaged – this is inevitable

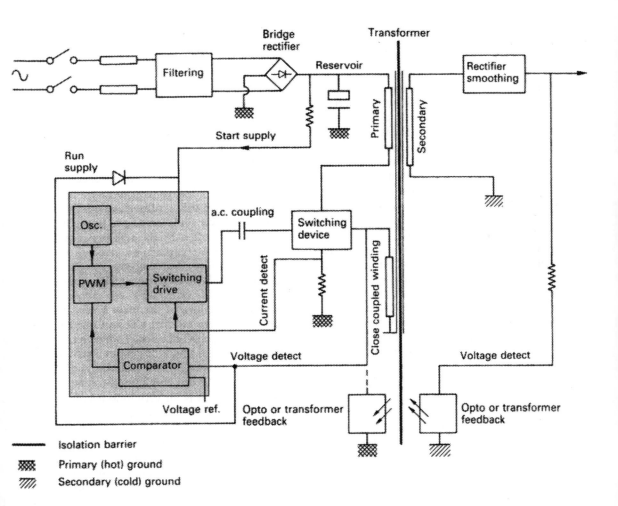

Figure 10.4   *Representative block diagrams of oscillator power supply configurations. All shaded items can be found in a single IC package, although the switching device is usually separate. Obviously, more than one secondary can be used.*

with d.c.-coupled circuits. Other faults in the primary can lead to regulation faults. Checks on the resistors in the feedback and reference circuits will often highlight the problem. Where you have a primary circuit failure and a kit is available use it!

The third failure type is where the supply is shut down because of a problem on the secondary such as a short circuit. Carry out a d.c. resistance check from the cathodes of the secondary rectifiers to ground – anything less than 200 Ω (other than heater supplies) is suspect. Lifting the loads from each supply in turn is an alternative way of trying to get the power supply to run. Do not overlook the possibility of short-circuit rectifiers, reservoir capacitors or regulators on the secondaries.

Overloads are often characterised by a squealing noise from the primary. Short circuits on the secondary can lead to destruction of the switching device and, therefore, other primary components.

In many cases, a secondary short, will lead to the PSU pumping – cyclically starting, shutting down and starting again. It is not always safe to assume that by lifting secondary loads in turn that when you get the right one, the pumping will start. Some units will pump with an insufficient load! When measuring the secondary that provides the 5 V supply, it can often read as low as 20 Ω due to the multiplicity of feeds from it. Do not be misled by this.

One must also remember that a power supply that is pumping – i.e. thinking it has an overload – may actually not have an overload but a problem in detecting one! A case of 'Who's guarding the guard!' The precise method of over-current (usually, but maybe over-voltage) detection will determine how this might happen or how likely it is to happen. Figure 10.5 is a good example. The 0.22 Ω resistor, R13 in the emitter circuit of Q1 the switching device, is used to sample current. By measuring the voltage drop across it, the circuit can (using Ohm's law) determine the entire current flowing in the power supply. If this current increases to an illegal amount, the drive from the IC will be removed (and then restarted and removed, etc., causing pumping). The value of this resistor is thus highly critical. If it were to rise in value by even a minute amount, you will end up with a pumping power supply. As stated earlier, such a component is likely to go open circuit if the switching device failed short circuit. Careful measurement then out of circuit with an accurate,

digital meter is necessary as is replacement with exactly the correct device.

Other primary circuit faults

Whilst being dead or pumping are the most likely failures that a primary circuit will cause, there are others. With the nature of feedback for regulation from the secondary see Fig. 10.6 as an example where an opto isolator U2 is used sampling secondary output and feeding back to control drive in the primary, if the feedback becomes modulated or disturbed by a secondary problem, a noise – usually a whistle – from the primary will result. Similarly if, as in Fig. 10.5 the primary sources feedback via the primary – here with a separate primary winding, a lack of smoothing would give the same result. Here C11 fits the frame.

Effects of hum and ripple on supplies

The usual victim of high temperatures in the PSU is any electrolytic capacitor, and as we shall see, the symptoms can be various. When a unit has run for a long time and then is unplugged for whatever reason, the cooling down can cause these capacitors to drop in value quite significantly and so it is that many new symptoms appear when the unit is repowered – often by the repairer in a workshop. Naturally as the capacitor's value drops, so does its effectiveness as a reservoir. It will depend on whether the PSU is linear or switched mode as to what kind of symptoms are attained. Linear supplies or failure of the main reservoir capacitor after the bridge rectifier will give rise to likely 100 Hz (or maybe 50 Hz if half wave rectifiers used) hum on rails and thus hum bars on vision, possibly upsetting field sync and low frequency hum on audio are the kinds of symptoms to be expected – this is all very long established failure experience. If we refer to the circuits in Figs 10.2 and 10.3, we see examples of where these problems would be the result of electrolytic failure. Similarly if C7 in Fig. 10.5 were to lose capacitance, the same symptoms may begin to occur despite this being a switched mode supply. However, in this case it would probably soon lead to primary circuit failure or indeed this capacitor with such a high voltage across it causing the mains input fuse to blow.

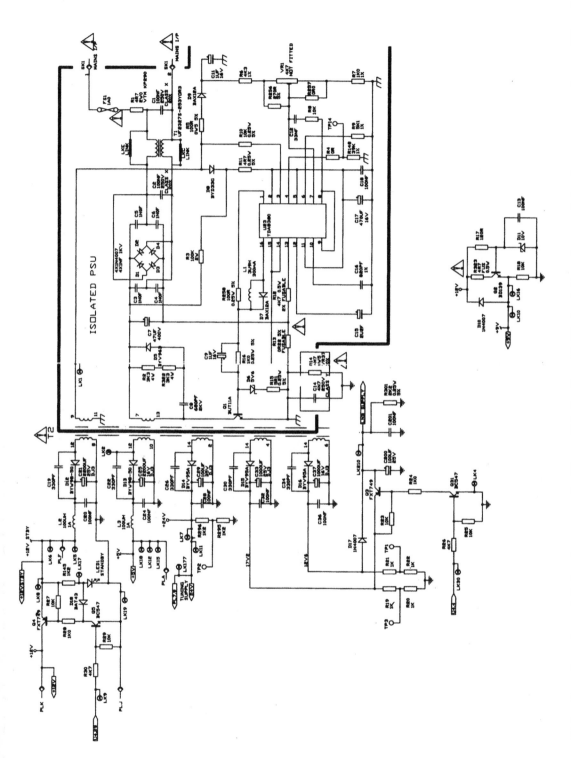

Figure 10.5 *An early receiver switched mode power supply. Many will be familiar with this circuit. Its rather open design demonstrates many features discussed in the text. A primary repair kit does exist for this design (Pace)*

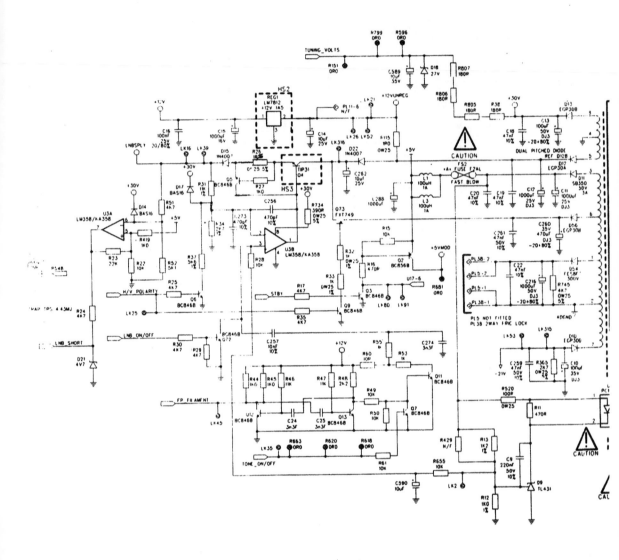

Figure 10.6 *A higher power switched mode supply using a FET rather than bi-polar switching device. Also evidenced is the inclusion of specific 'LNB SHORT' circuitry which not only protects the supply but alerts the user via an on-screen display rather than simply shutting the unit down (Pace)*

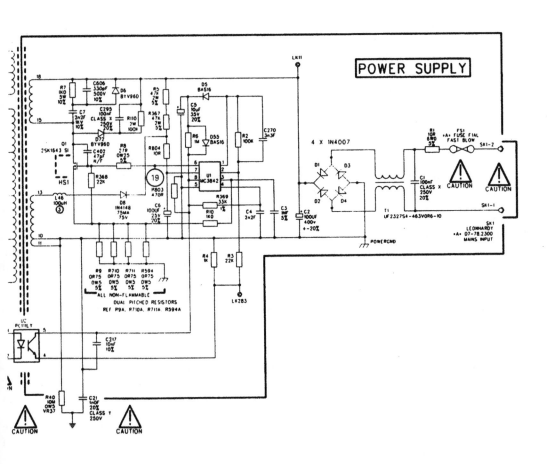

POWER SUPPLY

THIS CIRCUIT HAS BEEN SPLIT TO FACILITATE PRINTING

How to check

Obviously a 'scope across each capacitor would show the hum. Common sense then comes into play – a 47 µF capacitor with 330 V across it d.c. should not be considered faulty if there is 2 V p-p of hum present. However, in fault conditions you will see possibly 150 V p-p of hum there! Similarly an unregulated 12 V supply with 4 V p-p is in trouble but with 50 mV it isn't. It is easy to replace the capacitor for a definitive test – the traditional policy of tagging a replacement across the existing one (in parallel) on the print side of the PCB is perhaps best avoided these days – at least while the unit is powered! Modern print can be extremely intolerant.

As already stated, electrolytics tend to fail when cold and so judicious use of a hairdryer on suspect will, by heating it up, alter if not clear a fault. Similarly application of a chemical freezer will restore the symptom. One needs to be careful here as more than one component can be affected and extremes of temperature can take the component outside its normal operating thermal range anyway. This can all lead you a merry dance.

If you see electrolytics where the plastic sleeving around their body has shrivelled up due to heat, it is a fair bet that you have found your fault. All such capacitors should therefore be replaced. Larger value capacitors, especially where high voltages are concerned will 'rot' – i.e. a discharge will be apparent between their legs and one leg will often have corroded away. With mains reservoir capacitors, it is often the case that electrolyte fluid will leak from them.

Noise on switched mode supplies

Naturally, due to their mode of operation, a switched mode PSU gives rise to noise on supplies at a high frequency rather than the 50/100 Hz of linear supplies. In turn, should this noise reach a level that causes problems (i.e. when reservoir capacitors fail) the symptoms it produces are very different. Linear supplies produce ripples that are of a frequency that coincides with field scan rate and is within the audio range. Switched modes run at several kHz at least, therefore any symptoms on vision are likely to be more line frequency related. Different designs tend to have different symptoms but there are some distinct things to look for that will indicate a power supply fault:

- Lines, usually thin running through vision either horizontally or diagonally, and consisting of flashing or sparkling dots or smaller, bright lines.
- Dot patterns apparently randomly over vision. These dots can often be of relatively long duration and thus appear like sparklies. A known good signal or another receiver will instantly confirm but the clue is that the noise caused by the PSU will be less bright than a sparkly.
- Audible high pitched whistle from power supply area.

In any of these cases, eliminate the PSU at an early stage. If in doubt, then replace all electrolytics – this effort is seldom wasted!

Hum or noise on regulated supplies

With any hum or noise on a supply, if it is present on a regulated or stabilised supply, then obviously the reservoir on that regulated supply may be at fault. However, it pays dividends to 'scope the supply prior to the regulator itself where you will often find levels of noise or hum that the regulator simply can't cope with - the fault is thus with the reservoir on the unregulated input to the supply regulator.

Other electrolytic failures

Particularly in switched mode PSUs, electrolytics are not only used as reservoirs. As can be seen from Fig. 10.5, it is very common to find the base drive (for bipolar switching devices) is coupled via an electrolytic. This capacitor – here C9 – is a prime candidate in any circuit to 'dry up' and as its value falls so does its effectiveness at passing the drive signal undistorted. Favourite failure is for the switch-off portion of the drive to be attenuated and thus the transistor overheats and fails, usually short circuit. It is wise to replace an electrolytic in this position whenever you work on a unit with a switched mode PSU. It will save embarrassing recalls. Remember that electrolytics are worse from cold and so when the unit is returned following an unconnected repair and plugged in, it may well be slow to come on, or worse, blow the fuse as the switching transistor fails! Looking at the other end of this, it could happen to you when you first power it having

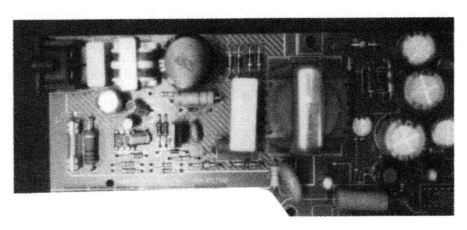

Figure 10.7 *A photograph showing the power supply area of the main PCB of a typical mainstream satellite receiver. Note the mains lead socket, collection of secondary capacitors and close proximity of all components generally*

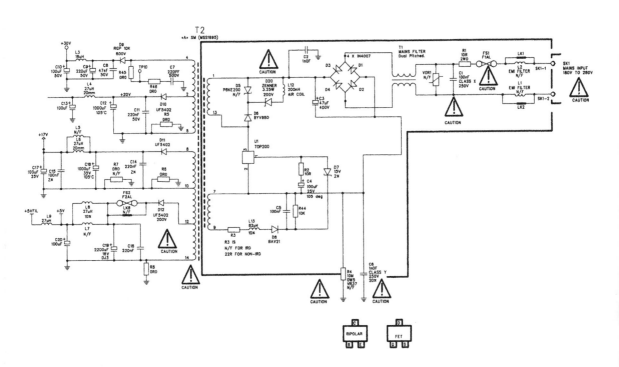

Figure 10.8 *A demonstration of current technology and how simple a primary can be made in a basic, lower power requirement design. The entire oscillator, drive and switching device circuits are contained in the three-legged IC enclosure (Pace)*

107

received the unit for repair. As such it is good practice to replace suspect electrolytics before even powering the unit, whatever the symptom may be, to avoid extra complications.

Stand-by switching

A d.c. control line from the microprocessor will, usually via a transistor or a relay or within a regulator i.c. switch on certain supplies for normal running. Conversely these will be removed in stand-by. With many satellite receivers, however, there is little that is removed during stand-by. Certain designs remove power to the LNB but most don't – for the good reason of temperature (and thus local oscillator) stability. If supplies don't come up when powered on from stand-by, then elementary tests on the switching line from the system controller followed by the switching transistor or relay should soon highlight the problem. It is possible that failing to come out of stand-by is a protection measure.

11

SYSTEM CONTROL CIRCUITS

The system control circuit is at the heart of any satellite receiver's operation. Its degrees of complexity and integration will depend on the age, manufacturer and features of the receiver. It will almost certainly be based around at least one microprocessor. One will encounter, in satellite receivers particularly, the inclusion of an emulation PCB containing various ICs in place of a single microprocessor that may be expected. See Fig. 11.1 – do not be misled by appearances. The role of system control is of co-ordination, control, synchronisation, monitoring, interfacing and protecting. This ensures that all circuits do what is required, when and protect in the event of misoperation. The role of protector means that symptoms can be very misleading for the service engineer – a receiver stuck in stand-by may have one thousand and one potential causes; a more dramatic scenario – say a smoking component – would be easier to find but at what potential cost to the unit? In a satellite receiver capacity, it will undertake roles common to any piece of consumer electronics – tuning control, keyscan and infrared receiver for user controls and remote control, display drives and menu generation, power switching, etc., but some specific roles may include polarisation switching, descrambling control, de-emphasis switching, etc. The principles and techniques are generic, one just has to appreciate their implementation. Throughout the book we have noted the effect and presence of system control in the area of discussion. This chapter covers the servicing of all facets of system controllers. It is very important to make careful

Figure 11.1   *An emulation PCB in place of a single microprocessor device*

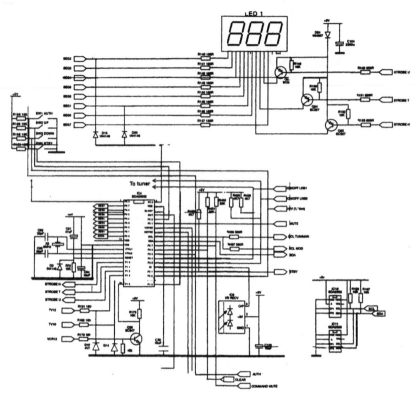

Figure 11.2 *A modern system control circuit illustrating various points in the text (Grove Farm Publications)*

checks when fault finding any such circuit otherwise unnecessary replacement of expensive microprocessor will follow. Figure 11.2 gives us a circuit to consider throughout the following.

Fault finding system control circuits

Total failure of a unit to operate can have two main causes: a power supply fault or a system control fault. Of course, due to the circuits' complexity this is only one of many symptoms that a syscon fault can cause. Here we shall deal with the possible reasons for this 'total' failure and how to determine whether it is indeed a syscon fault, as opposed to a power supply problem.

We discussed the action of a syscon circuit in switching a PSU on or off in the previous chapter. First examine all the output lines from the power supply to check for presence and correct level –

and also the condition of any switch line from the syscon. If this is incorrect any switched supplies in the PSU disappear but unswitched lines should, of course, be present. If they are, a syscon fault is likely to be present.

Three checks should be made on the system controller in this situation. First, are the supplies to the controller present and correct? If not, and so long as you have previously confirmed that all unswitched outputs are present at the power supply (from which the controller itself is powered, of course) the problem, if not due to excessive loading in the controller (lift supply pin(s) to confirm) will most likely be due to failure of any links or regulation circuitry between controller and PSU.

The second check should be on the controller's clock oscillator: for the presence of correct amplitude and frequency clock signals, whose failure is most commonly due to a faulty or dry-jointed crystal. Dry joints often develop because crystals usually have thinner legs than other components while the holes in PC boards are usually the same size for all components. A faulty Xtal, as it is commonly abbreviated, can sometimes be provoked into life with a sharp flick of the finger on its case. Be careful where you measure the Xtal waveform. Use a ×10 'scope probe applied to a recommended test point. Testing straight across the Xtal may possibly destroy it and the processor. At the least it can 'pull down' any clock signal present to give misleading information.

The third check should be on the RESET pulse for the processor: for shape, amplitude and duration. Many faults can be caused by a missing or permanent reset pulse. Sometimes a discrete reset circuit is used but often a separate, small IC generates the pulse from a d.c. supply.

In Fig. 11.2 we can see our microprocessor, IC4 with its 8 MHz Xtal, X2 on pins 12 and 13. Reset is applied on pin 14 courtesy of C61, D3 and R72. Supply is present on pin 11.

If a power-switching line to the power supply from system control is not operating correctly (usually not allowing the supplies to come on) one needs to look at system controller inputs. The likely causes, aside from microprocessor failure, are that a timer input line is activated – check user controls first – or a protection input is activating system control to switch off the power supply.

It is very common to find, most predominantly in system control circuits, what can be termed digital transistors. These devices are often drawn as a transistor in a square box rather than the traditional circle and showing the internal bias resistances between the three connections. This means that when checking devices the readings will be odd when compared with normal transistors. The values of the resistors can be 10 K or 47 K, for example, and the type number will convey this detail when you are familiar with the devices. IE DTA devices have one resistance value, DTC another, etc. Circuit references are sometimes not as for normal transistors. For example, where $Q \dots n$ is used for transistors then $QR \dots n$ will be used for these devices.

Lock-out

A noisy feed voltage to a microprocessor causes problems. Indeed, if mains- or signal-borne noise is experienced, such as during a storm, mains-supply variation or momentary breakdown, the microprocessor circuit can suffer. In most cases this problem is temporary – a *lockout* which is cured simply by disconnecting the unit from the mains for a few minutes and then reconnecting. If, then, a unit received for repair with the complaint of 'dead' works OK on test this is a strong possibility. This is a very common problem with Videocrypt decoders of earlier design.

Test programs

Many microprocessor-controlled units have a *test program* or *service mode*, usually accessed by pressing a combination of user buttons with or without an internal switch or link. These test programs can greatly aid diagnosis.

They may allow you to test certain circuits, for example a fluorescent display or a memory circuit, to the extent that the microprocessor will switch on all sections of the display at once, or provide a code on a display to advise you. These codes are used in conjunction with a 'look-up table' in the service manual which advises you of the expected results. Another indication could be a fault code. Here, in the event of a fault symptom, the test mode can be entered and a fault code read off, subject to these parts of the circuit working correctly. By no means is the engineer made redundant by this system, which provides invaluable assistance where a unit is 'just dead', and traditional fault finding thus made impossible. The information usually tells you the area where the fault occurred.

Input problems

When investigating syscon problems it is best, to begin with, to consider the syscon as a 'black box'. First check inputs and outputs. As the outputs are input-dependent, check the inputs first. If a receiver will not step channels up, check that the key input to the micro changes state upon pressing the key. (Also confirm that there is IR data from the receiver to the micro. If it is a remote model, failure to change via both methods would clearly indicate an output problem rather than an input one.) If so, look for an output from the micro to drive the tuning system. Where buses are involved these checks will involve looking for data changes on the bus when keys are pressed.

If all relevant inputs appear OK then you may have an output problem (see below). It is important to check an input, such as a key switch at source and again at the micro, but bear in mind that there may be intervening circuitry – inverters or input expanders, for example. Where a number of system control inputs are required and the micro has insufficient ports, an external input multiplexer IC will be used and the micro will address it to read all of its inputs which are latched into its memory.

Where inputs are d.c. switching lines they can have various different states. The obvious high or low states (usually approximately 5 V or 0 V) can often be supplemented by a half-way (2.5 V) state – tri-state switching. This naturally allows three modes to be indicated on a single d.c. line. Service manuals will advise what condition applies to each state. Careful d.c. checks need to be made, therefore, and suspicion should be aroused should you find a 4 V reading on a tri-state line or 2.7 V one on a bi-state line.

If the d.c. levels aren't correct we should start by looking at the basics. Are the d.c. supplies to the whole circuit correct in level and free from hum and hash? If not, look at power supply problems. If only one line is affected and it is low we need to check for correct d.c. pull-up or excess loading on the lines. Pull-up resistors can go open-circuit and thus a line will not be able to go high. Note that some designs use resistors to pull down to ground and so the opposite scenario applies.

When considering excess loading look at faulty ICs on the line, or protection devices such as Zener diodes used to prevent the line going above 5 V – they can go leaky or short-circuit.

When an incorrect d.c. level leads you to the other end of the line, away from the IC, consider likelihoods. Is a switch or mechanical interface used on the line? If so look at this first.

Keyboard inputs

The age of a receiver will be the determining factor as to what size and form of user input keys are present. Many receivers have no on-board controls at all, relying entirely on remote handset operation – a weakness perhaps. Others have minimal 'emergency controls' such as stand-by and channel step. Others, perhaps with no

remote operation, have large numbers of controls, maybe even mechanical direct switches which do not involve system control but simply appear in their respective circuits – a tuning thumb wheel for each preset channel, for example.

Later design use A–D conversion-key switches, connected to an A–D converter via a resistive ladder network, one section of which, when a key is pressed, forms a potential divider at the input to the converter. This applies a specific voltage to the input.

The modern way of arranging the user input is to use a keyscan system. Here a strobe signal (S0 to S7) scans the key network; when a key is pressed the microprocessor can determine exactly which one by virtue of the fact that the strobe outputs from 'keyscan' have different timings so that the inputs on 'pulse return' (K0 to K3) are unique for each key. A relatively new form of input key is the dial. Here a key matrix is formed in a circle pattern of contacts which are closed by the user moving a dial on the control panel. This technique is used for multi-purpose controls. For example, the dial can be configured by the user to provide channel change or volume control dependant on mode.

Output problems

Once the correct input conditions have been confirmed there are two main causes of incorrect outputs. First there is the possibility of a fault in the system controller or any ancillary switching circuitry which we will investigate below, but still considering the syscon as a block for the moment, there is also the possibility that a syscon line not changing state is due to a fault in the circuit to which it is connected, loading it excessively. Lift the load from the line to confirm this.

Incorrect output

Having eliminated all other possible causes of an incorrect output from a system control circuit we can now break down the 'black box' and examine the make-up of its circuit. Simpler circuits use the outputs coming direct from the main controller/microprocessor. Here, if the output is incorrect, we can confidently replace this device and the fault should be cleared – if not something covered above has been missed.

More often than not, however with the increasing range of features on modern units the output from the microprocessor is further switched by other related and dependent signals. This calls for a tracing of the errant signal backwards. We know that it is incorrect at its destination but is it correct at the input to the prior switching? If so we need to investigate, in the same manner, the switching line controlling the output. If we then find an incorrect output from the micro we must return to checking the relevant inputs for this output. It is a very common trap to fall into, where all thoughts are on the micro, to replace it without thinking further.

Often the output from the microprocessor is conveyed by a data bus.

Data buses

A slightly less helpful set of circumstances arise when two microprocessors or a processor and such ancillary devices as port expanders or memories are used together. Here the situation often arises where the input is to one whilst the output is from another. The two will be linked via a *bus* and it's only possible to confirm whether an input or output fault is present between the two if the manufacturer makes available the format of the data on the bus. Even then it can be difficult to display and interpret this data on a conventional oscilloscope – a digital storage 'scope greatly helps. Failing this, a check on the bus should reveal a movement or change in the data as the input (for example a user key) is actioned. If it does not, the input processor is likely to be faulty. If it does, and no output is present from the second processor, the latter is likely to be at fault but a degree of good fortune is required with this type of fault as well as the normal amounts of skill and brilliance!

If a unit is 'dead', or sections are inoperative due to a system control fault, it could be due to no bus activity where one is employed. A d.c. 'pull-up' is required for the bus lines to work and this is usually obtained via a small resistor to (usually) a 5 V rail. These buses are often protected by 5.1 V zener diodes which can go short-circuit to pull the bus down. Oscilloscope and d.c. checks reveal these problems.

Memory circuits

Modern designs rely on small memory chips to store tuning information and preset levels in virtually every piece of equipment. Faults usually boil down to an inability to store information. Sometimes the memory is inside the main microprocessor, in which case it will require replacement if the relevant supplies are present. Otherwise there is another possible problem: that of incorrect addressing from the microprocessor.

The supplies to the memory can be the cause of a problem. Where the memory is used to store tuning information check the tuning voltage supply carefully for correct level and regulation. A low level here may not appear to affect the tuning process, although the tuning point maybe wrong in correspondence with the scale, but it can upset the memory writing process. Often there's a 'writing' voltage of about 25 V, plus or minus. If this is present, being switched in only when storing takes place in some designs, and assuming that the correct procedure is being used by the operator to store the data, the memory IC, whatever form it takes, should be replaced. Inability to store information when the unit is not powered may also be due to a faulty memory, but certain designs rely on a 'back-up' memory supply in the form of a battery or large capacitor: this should be checked before condemning the memory IC itself. Where lithium batteries are used take great care when handling them – in certain conditions they can *explode*. In general, they are best left alone: certainly do not apply heat or a low-impedance meter to them.

Where NiCd batteries are used, do check the operation of any charging circuit that is used.

Displays

Another use commonly made of microprocessors is to drive a display or displays, either directly or via a driver circuit. Several different types of display are used. Their applications range from power on indication to audio level, channel and signal strength displays. See Fig. 11.3.

One of the most complex displays, capable of conveying relatively large amounts of information is the *fluorescent display* or *digitron*. This is a vacuum-tube arrangement and is driven via its grid or cathode. By their very nature display

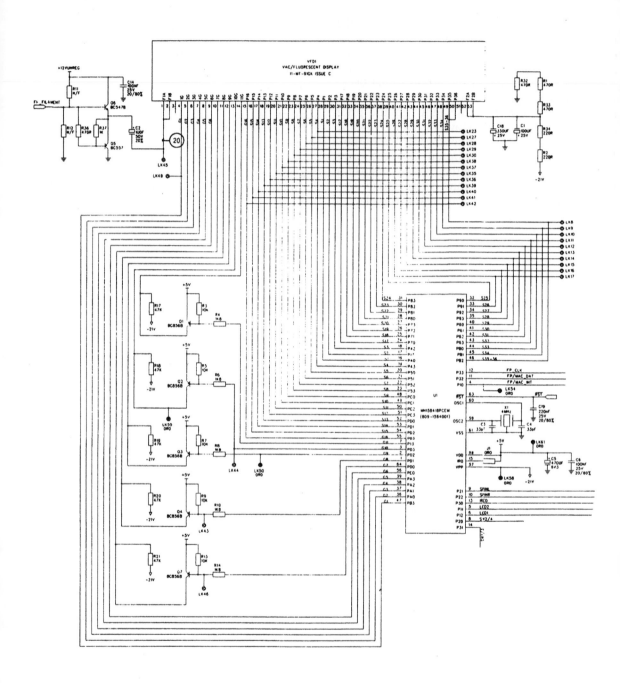

Figure 11.3 *A vacuum fluorescent display circuit illustrating the various points for testing (Pace)*

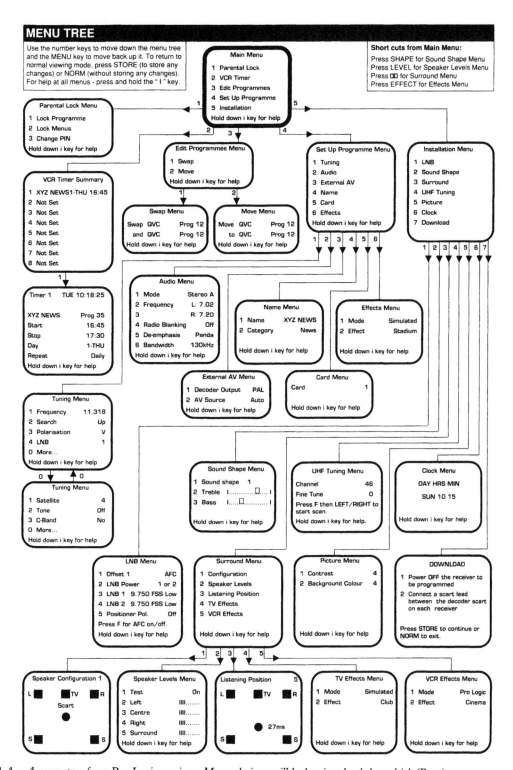

Figure 11.4   *A menu tree for a Pro-Logic receiver. Many designs will be less involved than this! (Pace)*

circuits are very reliable and any faults that do arise are often the result of damage, but of all the types used the digitron is the most common type to need attention. Heater supplies can be high frequency a.c. from a small oscillator circuit and, therefore, use of a 'scope is a good idea in checking level and presence. Check also, if necessary, the continuity of the tube's heater. In the case of damage the most likely result is that the tube has lost its vacuum due to a crack in the glass, usually at its weakest point: the glass nipple on its side where the envelope is sealed.

The other, more common affliction of the fluorescent display is that of low emission causing a weak, mottled glow with uneven segment illumination. The symptom is reminiscent of 'silvering' in old monochrome picture tubes, and replacement is the only cure. Check the supply to the heater carefully before condemning the display itself though. Replacements can be expensive and the symptom does not affect the operation of a unit, but replacing a 'tired' display can add value to a unit being refurbished for resale, for example.

LCD (liquid crystal displays) are rather less common. Faults here are rare indeed, usually taking the form of no display or missing segments. The cause is likely to be the driver device, but especially where segments go missing attention should first be directed to the drive signal conductors from the PC board to the layers of the diplay 'sandwich' – usually carbon/rubber affairs with contacts on the glass of the display. These can become warped or damaged resulting in loss of drive.

Dirt or dust can get grapped between the layers of the sandwich to cause problems: dismantling and blowing clean will cure this but handle the displays very carefully.

LED displays are used for level indication, most commonly in the form of a 'chaser' configuration where the row of LEDs light up to an extent reflecting the level of the signal reproduced. The scale is in dB and at the 0 dB point the LEDs tend to change in colour from yellow/green to red.

Problems are usually due to driver ICs or failure of the LEDs themselves. Often a faulty array has to be replaced as an entirety.

Most modern receivers incorporate on screen displays or menus as their main form of information display (Fig. 11.4). Graphics generation may be in the system control microprocessor or in a dedicated display IC. The requirement will be for H and V sync and the generator will provide blanking and data outputs to insert the graphics. If colour is involved, then a chroma sub carrier oscillator will also need to be implemented. All of these analogue signals then are valid test points in the event of a problem.

Infra-red remote control systems

This facility is almost obligatory on receivers and positioners. The system control circuit is extended to encompass an infra-red receiver as in Fig. 11.2. This feeds a data signal to an input of the microprocessor. Here the receiver is a can type module – older designs may well be discrete transistor amplifiers or a circuit (based on an IC) mounted on the PCB. Whatever the extent of the circuit, it will typically have three connections – supply (5 or 12 V), ground and data out.

The remote receiver can usually be quickly identified and located within a unit due to the facts that it is usually heavily screened and also often mounted away from the rest of the circuit boards. It must of course also be visible to the remote-control beam through the front of the unit.

Handset faults

Despite their simple makeup almost as much can be said about repairing handsets as about many other more complex circuits. The author once wrote a five-page article on the subject which only scratched the surface! To be practical in a book like this we need not list fault symptoms and causes, but simply cover the reasons for failure – this leads to easier fault tracing. We need to remember that handsets are used heavily; have liquids spilled into them; get dropped; have many small holes into which a great variety of foreign items can fall; and use batteries! Therefore look for spillage, dry joints (particularly around the IR LEDS), PC board cracks and breaks, weak batteries, poor battery contact and worn button contacts. Battery covers are common victims, usually losing their tension and falling off, causing the batteries to make poor contact. Conductive silver paint (available from various service aids suppliers) can be used for many repairs from bridging print breaks to 'retreading' worn rubber contacts.

Incorrect or no output, or excessive battery drain (if not caused by the more likely spillage) are usually due to failure of the IC. Broken legs on the crystal or ceramic resonator are another common problem. Spillage should be dealt with using the methods described in Chapter 15.

The cost of replacement handsets has, on average, declined in recent years and the trend among manufacturers is toward supplying no spares save, say, the battery cover. Pattern handsets from third-party suppliers can be amazingly cheap, but again with some restrictions and they are often aesthetically inferior.

Pre-programmed 'universal' handsets are very popular but often do not have the required codes for older or less well known receivers. Similarly – as with TV and VCR – they do not support more than the very basic commands and so may prevent tuning for example.

Receiver faults

In practice most remote control faults are due to problems with the handset or, in the case where one unit receives the signal and then relays it to other units in the system, to interconnections. Ensure that all connections are good.

In the event of failure to respond to remote commands, first check that the front-panel controls work OK. If so confirm that the d.c. supply to the remote receiver/pre-amp is intact and free of noise, ripple and hash. Next check for a data output when an IR input is present (function key held down on handset and handset pointed at receiver). Few manuals provide information on the type and level of signal to be expected and so experience comes into play. Check for a peak-to-peak level equal to the IR amplifier supply voltage – check that the supply voltage is of the correct level and clean. Look also for changes in the pulse width between different commands. At this point in fault tracing the *presence* of data is a sufficient prompt to proceed and check that it is reaching the system controller. Once this is confirmed we apply the same logic, described earlier, as for any other input/output fault. One unusual problem is where a button on the unit itself is faulty causing it to be permanently on, thereby 'locking-out' remote commands, and all other inputs.

The nasty side of a fault like this arises when there is *incorrect* data, which will be ignored by the system controller. The obvious course of action, if data is present at the input but no output is forthcoming, is to replace the microprocessor, but in a few cases this will not solve the problem, confirming that the data is incorrect. Luckily the circuits involved are not complex, usually involving an IC and a few passive components, because little in the way of fault finding can be done in these circumstances. The ideal fault-finding process is to make comparisons with an identical unit if possible – before getting deeply involved in this fault do ensure that the fault isn't in the handset by trying another one, this will save you hours of heartache!

The more common symptom caused by incorrect data is that of remote-control keys doing the wrong things. In older systems where a coil is provided for adjustment setting-up may be all that is required, the fault being due to component values changing with age. Ideally of course you would find the out-of-tolerance component, perhaps a capacitor. In other cases the fault is usually due to a faulty amplifier/decoder IC or associated capacitors. Thankfully these latter faults are not very common.

A perennial problem is that of random 'remote' information reaching the controller, be it from a faulty handset (spillage, etc.) or noise being generated in the receiver. The simple test, if the complaint is of random function jumping, etc., is to disconnect the receiver's output from the controller.

Finally, before spending hours on a receiver fault, do check the price and availability of the complete receiver if it is separate from the rest of the circuit. They are often surprisingly cheap to buy – not the most satisfying repair but the most cost-effective.

Problems

There are one or two features of system control that may cause you to delve into apparent faults which in actual fact are not faults. Many satellite receivers have timer circuits which allow them to be preset like a VCR – if set to a timer mode, the receiver may well appear to be stuck on stand-by. Certain satellite receivers now incorporate timers with VideoPlusTM, a system by Gemstar Corporation which allows a numeric code (from a TV listings guide) to be entered instead of all the date/ channel/time information conventionally required for timer setting. This system is now widely used on VCRs. Similarly many modern VCRs will, via an infra-red transmitter, control the satellite receiver – i.e. switch on and set channel at prescribed time – to allow timer recordings to be made. This, naturally, requires the VCR to be set up to the correct set of pre-programmed infra-red codes for the receiver in question. It is a fact that many satellite receivers are unsupported by codes.

Software

Similarly many circuits include child-lock/parental control or security code locks. It is human nature that these codes will be forgotten and thus access to some or all features prevented – this may be certain channels or menus, for example. Most (good) designs have the facility of a software reset so that the PIN (personal identification number) for the locks can be defaulted to say 1234 or 9999. A software reset may also be available to restore all settings (including tuning) to that programmed at the factory. Manufacturers' service manuals will advise, but do bear in mind that if you carry out such a reset, much tuning information and channel naming may require redoing!

Electronic variable resistors (EVRs)

This new technique has already found itself in TV, video and camcorder products and will undoubtedly find its way into all consumer electronics eventually as digital signal processing becomes more commonplace. The traditional variable resistor adjustments are no longer present, and to facilitate adjustment requires data communication with the control microprocessor and storage of the data in the unit's memory.

There are various ways of achieving this. Use of a service mode is the common link; once accessed, the unit's remote handset or a computer or special handset/jig are used. Each adjustment is accessed as an address and its value read out, altered and re-stored.

There are new possibilities now with interfacing PCs with receivers to provide diagnostic information and programmability. Already, via pin 8 of the scart socket, receivers can be used to program each other and various PC utilities and interfaces are available to enable rapid, block programming of channel information. Digital receivers have RS232 serial interfaces for the user and engineer. Software upgrades and diagnostics will be available via this port and fault finding will be less with a meter and 'scope and more with a PC. Service engineers really will have to change their methods.

12

R.F. MODULATORS AND AMPLIFIERS

These devices have been around in consumer electronics for a very long time now – predominantly in VCRs, but their development and thus servicing had been minimal until a few years ago when Pace decided to do something a bit different with theirs in their satellite receivers. It still begs the question: why did no one manage to do it sooner and why do many still not do it now?

What?

R.f. modulators take an audio and/or video signal and modulate it onto a r.f. carrier, in the UK a u.h.f. carrier, usually on a channel that is adjustable between 30 and 40. There have been channels in this area of the band that were not used for u.h.f. transmission (35–7) and so were suitable for these closed circuit applications. The advent of Channel 5 in the UK has used channels 35 and 37. With the possibility of the four existing UK terrestrial channels being in this area and C5 on 35 or 37, things could well get cramped. There must be at least two channel spacing between used channels (16 MHz) and a VCR sitting in the 30–40 band (not to mention possible games machines with non-adjustable outputs) would make life very difficult if you had to introduce a satellite receiver which, like VCRs, usually has (or had) a modulator capable of outputing between 30 and 40.

Moving the modulator's output range is not an answer as wherever you put it, with only 10 or so channels' capacity, it will always be in someone's way – certain VCR manufacturers do this and it creates more problems than it solves! The answer is to make your modulator broadband, and that is what Pace did. Not only that, but it is software controlled so that you can simply type in the channel number at which you wish it to operate. This means that you can place its output anywhere in the band where you don't have anything else. They also made the amplifier/modulator a

discrete part of the circuit – i.e. built on the main circuit board. This means no more easy (but expensive) module changes! see Fig. 12.1

Conventional modulators/amplifiers

The standard form has been to have a modulator, probably including the r.f. loopthrough amplifier, built into a modular can. Should a failure occur, the can would be replaced. Note should be made that there is in the UK a repair facility for r.f. amplifiers, modulators and tuners – refer to Appendix 3. This module then would have supplies, a ground, audio and video baseband feeds and possibly some option switching. Such an arrangement can be seen in Fig. 12.2.

Voltage supplies

The supply labelled 12 V STBY and connected to the pin marked +B (albeit dropped to 9 V) is present as long as the receiver is powered and this supply powers the r.f. amplifer which loops the terrestrial signals through the receiver. Refer to Fig. 12.3 where we can see that the u.h.f. aerial is looped through the satellite receiver first so that the output of it can be introduced prior to the VCR thus enabling satellite programmes to be recorded. If the satellite receiver's r.f. amplifier failed, then terrestrial signals would be severely attenuated.

Causes of this may be that the receiver is totally dead – it is amazing how many people will call you for snowy pictures not having realised that either the satellite receiver or indeed VCR is dead. Alternatively this supply may have gone missing. If no loopthrough is the only symptom suspect that the cause of the loss is very close to the module – a broken print track for example – as the stand-by

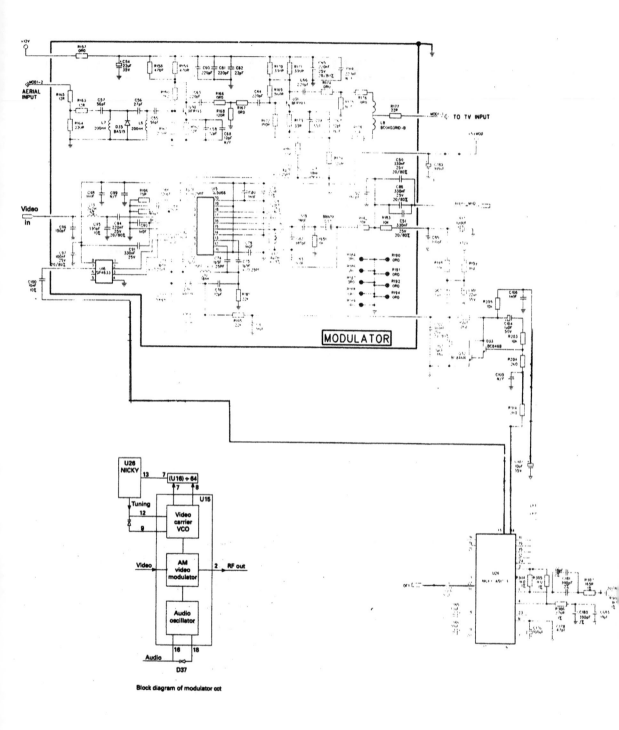

Figure 12.1 *Current discrete r.f. amplifier and modulator circuit. R.f. output is frequency synthesis tunable and the circuit is simply a screened part of the main PCB rather than a separate, replaceable module. A block diagram of the operation principle is also shown (Pace)*

120

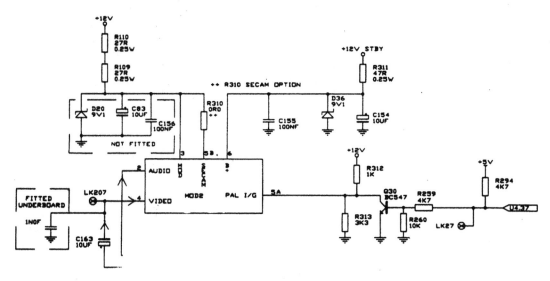

Figure 12.2   *The conventional format of an r.f. amplifier/modulator (Pace)*

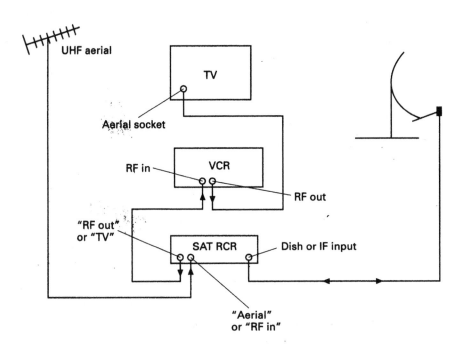

Figure 12.3   *The distribution of r.f. in a domestic installation*

supply will be used elsewhere as well and so if it were completely missing (from the PSU) there would be other symptoms.

The second supply, labelled +12V and connected to the pin marked 'MOD' will not be present in stand-by (usually) but is used to power the r.f. modulator section of the module. This is of course only necessary when the receiver is on, to enable the output to be viewed at r.f. If it were missing the receiver would work, viewed at scart, everything would be fine but there would be no r.f. output – loopthrough would also be fine. Do not overlook the possibility of a user option (via menu) to switch off the r.f. modulator.

Signal inputs

We obviously have audio and video feeds to the modulator section of the module. These will usually be d.c. isolated (capacitor coupled) and buffered from any baseband outputs from the receiver such as scart or phono connectors. As with all audio and video faults with receivers we should check to see if the symptoms are present at r.f. and baseband. If they are, carry on fault finding but if the fault is only present at r.f. then start looking at the modulator.

In doing so we can easily identify whether the problem is in the modulator itself or in the likely small amount of buffering circuitry between it and the take-off point for the baseband feed that is known to be OK. We will be looking for baseband level signals at the input to the modulator – i.e. 1V p-p video and a couple of hundred mV of audio. If the baseband feeds in are OK and supplies and ground are present and correct, then you have a fault internal to the modulator.

The usual kinds of faults that occur with r.f. modulators are: loss or severe distortion of vision (the vision may appear negative with sync corruption and over saturated chroma); audio can be distorted or missing. Remember to check these at baseband first though!

Options and switching

There are a number of parameters that a receiver may be able to switch in the amplifier/modulator. A couple of examples are present in Fig. 12.2. A d.c. control line (labelled 5A) switches the audio

modulation between 5.5 MHz and 6 MHz inter-carriers (the latter for UK, PAL I and the former for PAL B/G as in much of continental Europe). Were this to be inappropriately set, there would be no or buzzy distorted audio at r.f. only. Again baseband would be fine. Pin 5B is a manufacturer's option to enable a SECAM modulator as opposed to PAL.

Broadband modulators

Traditionally, the ten or so channel modulator's output frequency was adjusted by turning a variable capacitor accessible via the outside of the receiver. These were not really a problem but could be fiddly and were prone to damage if customers got near them! The Pace broadband modulator is electronically controlled using a varicap system. In Fig. 12.1, we can see within the boxed area (which remember is purely a screened part of the main PCB, not a separate module) a varactor diode, D36. A d.c. tuning voltage (0–24 V in this example) is applied to it via R195 from IC reference U26 which is actually named NICKY 3. This is a device specifically designed for frequency synthesis tuning of the satellite tuner, the audio and the r.f. modulator by Pace (an ASIC – Application Specific Integrated Circuit) so I suppose they can call it what they like!

Varying of the d.c. here moves the modulator's output in the same way as any tuning system – as encountered in TV and VCR. The details of PLLs and tuning systems are covered in Chapter 7. Clearly though, if you have no r.f. output and the modulator supply is present (here labelled +5VMOD) then this tuning voltage is an extra consideration beyond the conventional modulator checks already outlined. It is possible for the tuning to take the output above the normal u.h.f. range of 21–68. Failures have been experienced where there is an output but it can only be seen on TVs capable of tuning up into the area of channel 74! Any problems here should be addressed as with any other tuning system fault.

Setting r.f. outputs

In an ideal world, all that you need to do is to find a space in the u.h.f. spectrum where you can achieve two channel spacing either side of the

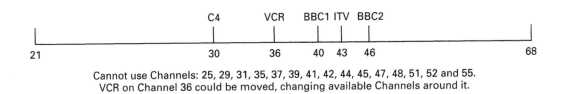

Cannot use Channels: 25, 29, 31, 35, 37, 39, 41, 42, 44, 45, 47, 48, 51, 52 and 55.
VCR on Channel 36 could be moved, changing available Channels around it.

Figure 12.4   *Spectrum of a local area showing available channels*

receiver's output and where you do not sit it upon $n + 5$ or $n + 9$ where n is the channel number of another channel, $n + 5$ being where interference from n's local oscillator may occur and $n + 9$ being where second image channel interference may occur. Similarly $n - 5$ should be avoided. See Fig. 12.4 where we can see a local u.h.f. spectrum and some of the no-go areas. All of this assumes that we have well balanced signals and that they are relatively strong and trouble free. Often, of course, this is very far from the case and so finding a clear space can be a real headache, especially without a broadband modulator and with unbalanced signals with multipath distortion! It may seem ludicrous but a spectrum analyser would be very useful for the installation of a VCR or satellite receiver in such instances!

Without such luxuries, one can make use of frequency synthesis TVs or VCRs. Use their displays to tell you what channels are currently being used, plan what you want to set the receiver to and then store that channel on a spare location on the TV or VCR. Move the r.f. output so that it then sits perfectly on the location. Those receivers with broadband, frequency synthesis modulators can be programmed as to where to go thus making life very much easier. Failing this, one can use a signal strength meter if you can dial in carriers. It is not desirable to set r.f. outputs off channel – i.e. between two channels – unless no alternative exists. Take note that carrier frequencies for u.h.f. channels are usually nnn.25 MHz, e.g. 743.25 MHz for channel 55.

Internal modulator failures

Where it is necessary to repair modulators – e.g. as in Fig. 12.1 or 12.5 – then the circuit is invariably based around an IC. Having established that our baseband feeds are reaching it, we need also to confirm that supplies and ground are present to it and that the oscillator is running at the correct frequency. Poor video response (smearing, etc.), poor sync or buzzy or low audio could all be due to misalignment although you would have to question whether this was caused by a component having aged, drifted or started to fail. With video modulation problems, check the d.c. level into the IC to see what it is sitting on. Clearly if this is wrong all manner of problems with black level or sync may occur. External components, here R197, set the video modulation index (to 80%) and so overloading or low contrast r.f. would suggest a problem here. In our example, Fig. 12.1, there is also a varicap tuning system for the audio carrier at D37 and so any loss of r.f. audio would warrant this extra check.

R.f. amplifiers

These are relatively simple pieces of electronics that can give some real problems – and not always failures. Likely failures first, though, poor r.f. loopthrough due to loss of supplies we have already covered. Some other possibilities involve

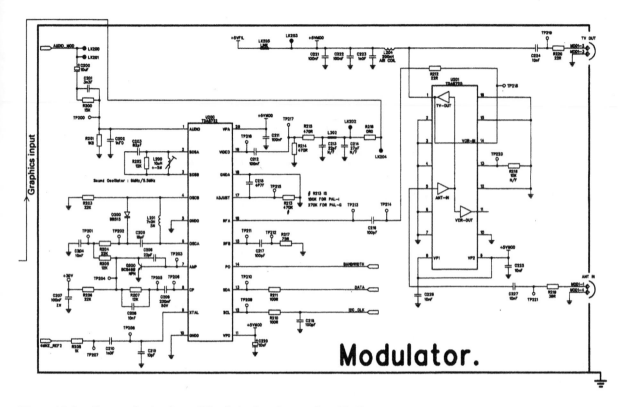

Figure 12.5 *Current, basic r.f. amplifier design based around an IC (Pace)*

failures within the amplifier circuit itself. In a can type device, this means a new or rebuilt can. With circuits built onto the PCB such as that shown in Fig. 12.1, the cause of the problem has to be found to component level. This example is a fairly conventional low noise amplifier using r.f. transistors. Any low gain would almost certainly be due to them failing. More recent designs have seen the r.f. amplifier within an IC. See Fig. 12.5, where U201 is the IC in question, U200 being the modulator IC.

The quality of design and build with some amplifiers (and modulators) can be very poor. Cost conscious mass production sees to that. R.f. amplifiers fitted to VCRs or satellite receivers have to be broadband to allow them to be used anywhere in a given market. They also tend to assume that they will be looping through only four or now five terrestrial channels. Like any r.f. amplifiers if they are broadband, their overall gain will not be as good as a grouped amplifier.

Furthermore, if you have more than one transmitter available to you, i.e. more than the usual four or five channels, then the r.f. amplifiers can start to perform rather badly. The gain will drop and by the time that it is also passing the r.f. modulated receiver output the results can be very poor indeed. Many modern units exhibit these problems. Similarly too high an r.f. input to many amplifiers can cause the output to appear low gain. If you suspect this, try attenuating the input with a variable attenuator. It may seem perverse when you are looking at a grainy output but it may just cure it!

This is all exasperated by the fact that by reducing the off-air terrestrial signal level like this you make it more prone to interference, and around an average satellite receiver there is plenty of that to be picked up. The two most likely areas are clock signals from the Videocrypt decoder or radiation from the switched mode power supply. Current European legislation lays down minimum standards for interference causing radiation from electrical equipment, but don't bank on that stopping any problems!

Dealing with interference pick-up

If you do have genuine problems with interference from satellite receivers or decoders, there are some basic things to do which will often cure problems. Do check with manufacturers as well, however, to see if any specific advice or modification exists for your problem – it will often be found to be the case.

Ensure that all r.f. interconnecting leads between receiver, VCR and TV are double screened (CT100), hand-made with soldered plugs. Ensure that the receiver case and aerial system are grounded and that the case of the receiver is grounded to the PCB secondary ground. Reduce the length of interconnecting leads to a minimum and ensure that there is good physical spacing between all units. If the receiver is in a cabinet or enclosure, line the roof of it with tin foil to add screening.

R.f. amplifier problems

Where design of the amplifier leads to u.h.f. problems, the only real answer is to improve the u.h.f. system to cope – add amplifiers prior to the receiver to give it the lift required. This will not always work. Otherwise filter out the weakest channels with a grouped filter leaving only the strongest group. Failing that, do not use r.f. connections through the receiver, instead rely on scart or other baseband connections. This may not always be possible due to availability of sockets or inability of VCR to record from an A/V input on timer.

Amplifier/modulator problems

The cheapened designs of the combined modules discussed above affects not only loopthrough but injection as well. The modulators are often not very well set up and not at all fussy about levels or sound/vision ratios. It is perfectly common to find sound 20 dB down on vision on receiver modulators and with many televisions, this will result in buzz on the sound. Imagine the potential for problems when you also consider the possible poor performance of the r.f. amplifer. Another problem is the interference that can be caused to adjacent TV channels by the huge sidebands produced by some modulators. Fine tuning of the output channel can take ages to clear patterning from apparently unconnected channels.

Fault finding in general

It is possible to easily identify these problems without a circuit diagram. As can be seen, there are few connections to an amplifier/modulator and they are invariably the same ones. A check with a 'scope and meter allows quick identification of these and therefore enable fault finding.

13

DECODERS AND DESCRAMBLERS

Here we are considering commonplace units – i.e. those involving Videocrypt and D2-MAC – which are not covered elsewhere. Naturally the UK reader will be far more interested in Videocrypt than MAC, conversely D2-MAC will be of more interest to the Scandinavian reader. In the UK, D2-MAC is not in widespread use as it is in much of Scandinavia but there is a healthy market among those who view services not intended for the UK by using so-called 'pirate' smart cards. Similarly there is a band of viewers who utilise the same type of device to view Videocrypt encrypted programmes such as those from BSkyB without having to pay for them.

Security

Due to the nature of the business – descramblers are usually used to provide encryption allied to obtaining subscriptions to services – information about how certain parts of such systems work is guarded. Equally, certain ICs which are used in the circuits of decoders are not generally available necessitating the return of the board to the manufacturer for repair. This is characteristic of Videocrypt which is licensed by Thomson consumer electronics. Life is not made easy for the service engineer in such circumstances but experience proves that failures in these restricted areas are few and far between. When they do occur, they are usually within older units and so there may well be scrap boards around from which to acquire ICs for repair. This will be far more acceptable to most engineers who hate admitting defeat! Paying customers will usually appreciate this uncharacteristic use of second-hand devices as opposed to paying for a unit to be returned to a manufacturer (although certain ones do provide an excellent service). To add to the difficulty of fault finding on Videocrypt circuits, the diagram provided is usually covered in broad bands of dark ink to

prevent it being photocopied and each one is serial numbered.

Decoder interfacing

Many problems can be caused by interfacing problems, especially when an unofficial decoder of some sort is being implemented. Many 'pirate' decoders are available to allow viewing of certain services outside of their intended territories and so it is that you may find a problem caused by a user incorrectly connecting one or indeed one being faulty or poorly designed.

Loss of vision or sound (if an audio decoder is used or audio is routed to the decoder) can often be due to interconnecting leads or the decoder being faulty. Some receiver designs loop all channels through the decoder connector whether they need to be or not. Some decoders loop back into the receiver (as with Videocrypt) but others do not – they have their own r.f. modulator, for example. Investigate signal routing to establish where the fault may lay. Interference on sound or vision of any channel may well be caused by a fitted decoder. An example recently encountered was of a brand new Astra system with hum on vision. The cause was the power supply in an external RTL5 decoder owned by the customer. It's linear supply had failed giving rise to ripple on supplies. This found its way back into the receiver. Disconnecting was sufficient to prove.

Where decoders seem not to work, check that the routing of the signal by the receiver is correctly set up. This will be done via the menu system if it has one. In most cases, each preset will be able to be routed independently. Other problems are often due to the wrong settings for video de-emphasis; external Videocrypt units have a de-emphasis switch which should be off – it gets switched on causing the scrambled image to be very smeary and there is no descrambling taking place.

Equally, some decoders (e.g. Videocrypt) require baseband video, others require clamped, composite video. Getting this wrong will result in incorrect – often intermittent – operation of the decoder.

Videocrypt

Figure 13.1 illustrates the layout of a Videocrypt decoder. In older designs this represented many ICs and much external circuitry. Modern IRDs (Integrated Receiver Decoders) have a much more integrated layout. Videocrypt works on the basis of taking a line of video, selecting one of 256 possible cut points and then cutting the line and

rotating the two pieces of video about each other. These cut points therefore need to be known at the descrambler in order to reconstruct the line correctly. This is achieved by means of an algorithm transmitted in the field blanking period (only the active part of the scan is scrambled) in conjunction with a decryption algorithm stored in the microprocessor of the smart card (or Viewing card as it is called by BSkyB). The decoder does not contain any secrecy algorithm – it serves only to validate the card and descramble the image. Also present in the blanking period is service information such as channel name which is used to produce on-screen graphics. Figure 13.2 shows the on-screen effect of a Videocrypted picture with no decoder used.

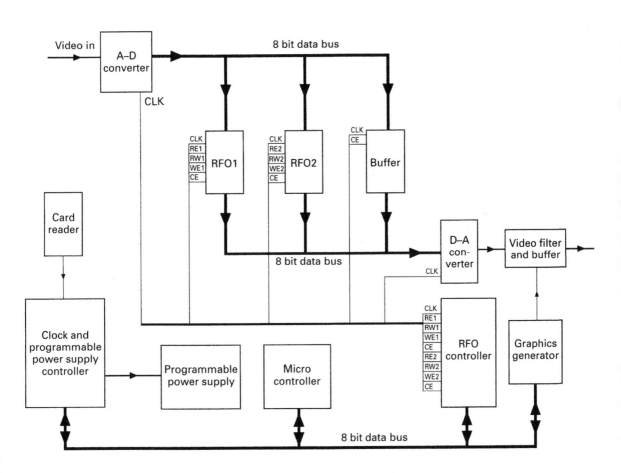

Figure 13.1   *Block diagram of a Videocrypt decoder*

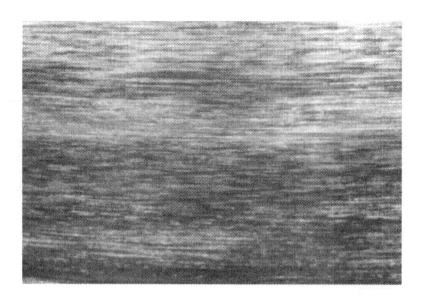

Figure 13.2   *A Videocrypt scrambled picture with no decoder attached*

Security process

The signal into the decoder has its control data fed to the smart card which is used to reconstruct data to select the appropriate 'seed'. This seed is then used by the decoder to find the correct cut points and descramble the picture. Incoming video is low-pass filtered and then analogue to digital converted for descrambling and then converted back to analogue video signal. Fault finding in this middle area therefore consists of scoping digital video data.

Pirate Videocrypt hacks

There is widely available information on a serial interface (not to mention pre-made PCBs) that can be inserted into the card slot of a Videocrypt decoder in place of a legitimate card and then connected to the serial port of a PC which then controls descrambling. This is under the control of certain software of course. This software is available from certain sources on the Internet and hackers provide it as soon as they have cracked the

latest efforts of the broadcaster. It is very interesting to note that in the past, this has only taken a few days following the release of a new generation of cards from, say BSkyB, but things have been very much quieter since the launch of the BSkyB series 10 card which is currently bing updated to series 11.

Videocrypt faults

When a card is read, power supplies have to be provided and so failures related to failure to read here should direct attention to supplies. The most common failure here, though, is the card reader itself. Poor contacts are common – tiny movements of the card will highlight problems. The insert switch in the slot often gives trouble resulting in a 'PLEASE INSERT CARD' message even though a card is present. Another common failure with card readers is when they split in half. The construction is usually two plastic halves heat-melted together. The lids pop off causing poor or no contact with the card. The most likely reason for this seems to be the use of pirate interfaces to fraudulently view services. The PCB that replaces the card is thicker and thus spreads the card reader until it breaks.

Another problem that can seem perplexing is that of card failure. Many symptoms which come down to not descrambling – or often randomly breaking into scrambled vision for a few seconds – or putting random messages up on the screen can be due to faulty cards. That is not to say that they are unreliable – many are blamed when not the cause of a problem. BSkyB will replace one that is faulty but will charge if it is lost or if multiple failures occur. If you are an ASA or authorised BSkyB dealer, then you will have a demonstration card which can be used for testing purposes – this or a replacement decoder is the best test, certainly in the field.

It is highly possible though that card failure has been precipitated by a fault within the decoder. Older designs require modification to work with the latest cards without damaging them. Check with the manufacturer for specific details if you have an older design damaging cards. Similarly noise on the supplies to the card reader will corrupt them – carry out 'scope checks on the PSU rails. The likely problem is, of course, reservoir capacitors. Never replace a card more than once without investigating the decoder.

Lock-ups

One problem that many designs of Videocrypt decoder are prone to is that of lock-up. The symptom is that it will not descramble and there are no messages on screen from the decoder (receiver generated messages may well be fine). Simply resetting the mains supply to the unit will usually cure this. If the symptom persists, ensure that supplies are present and correct before getting more deeply involved.

Vision faults

Transistor buffering stages are usually present (especially on older designs) at the video output stages of the decoder. They can give rise to many problems such as no or weak/negative vision. Routing an unscrambled signal through the decoder will add weight to this being the area of trouble if they are similarly afflicted. Simply linking out the input to the decoder to the output with a capacitor will bypass it and prove a point. Without a circuit, this type of problem can be easily investigated by scoping back from the video output. Sync corruption can also occur as a result of incorrect biasing of these buffer circuits – this will naturally have an effect on the on-screen messages as well as the picture video information.

Some rather strange faults can be caused by faults within the descrambling part of the Videocrypt decoder. One common one is where white dots appear randomly over the descrambled picture. Having ensured that the PSU rails are clean, the problem is likely to be the A–D converter. A similar problem can occur with black dots! The point to note is that these faults only appear when vision is routed via the decoder – try a clear signal to prove. The dots in question are not comet-shaped like sparklies. Do not confuse the two.

With all other video faults, look at supplies and oscillators first. Older stand-alone units (SVA1) used linear PSUs and reservoir capacitor failures are very common. This can cause all manner of problems including on-screen messages that are not normally seen.

Random operation of Videocrypt decoder

It is often the case that a channel will intermittently drop back to its scrambled state and then a message will appear – usually 'PLEASE WAIT' and then it will clear again. There are many possible causes. Make sure that the signal level is sufficient and that the video signal, especially the blanking periods are not corrupted. This may be as simple as weak signals – a dish off beam for instance – or if more channel specific, a mismatch. The simplest test is a replacement receiver and running the suspect one on a different feed. If the signals are OK, look at the user contrast control. These alter the video level prior to the decoder (if it is built-in) and if set too low will lead to insufficient video level reaching the decoder and thus misoperation. Set the level to at least half scale to be sure. It is a fact that Videocrypt signals do very occasionally simply do this for no reason as regards the receiving equipment. Similarly a temporary uplink problem (severe weather at uplink site) will lead to some problems. These are the minority cases though. We have already said that a faulty viewing card will also cause the problem.

Radiation

It is an unfortunate fact of life that Videocrypt decoders radiate interference. Modern designs are far better but older ones wipe out an off-air u.h.f. channel with no effort! The cause is radiation from the clock signals (several MHz, typically 17 or 28 MHz) used. The patterning is akin to adjacent channel interference – wavy lines or herringbone patterning on vision. Circuit design plays a large part in how little is radiated – long lengths of printed circuit carrying the signals are not ideal. Certain receiver designs have had modifications to

alleviate the problem but as with all forms of interference, there are certain general things that can be done.

Ensure that all grounds in the receiver are intact and that the r.f. sockets are bonded to the case. Replace all interconnecting leads with short length, double screened ones, hand-made (not pre-fabricated!), and with all plugs soldered on. If necessary, replace u.h.f. down-leads as well. Try to improve the u.h.f. signal level – if it is low, the TV and VCR AGC will be working hard and thus be more prone to collecting the radiation from the Videocrypt unit. It has to be accepted that some customers are rather less than pleased that they have to have work done on their terrestrial aerial system due to the introduction of a new satellite system!

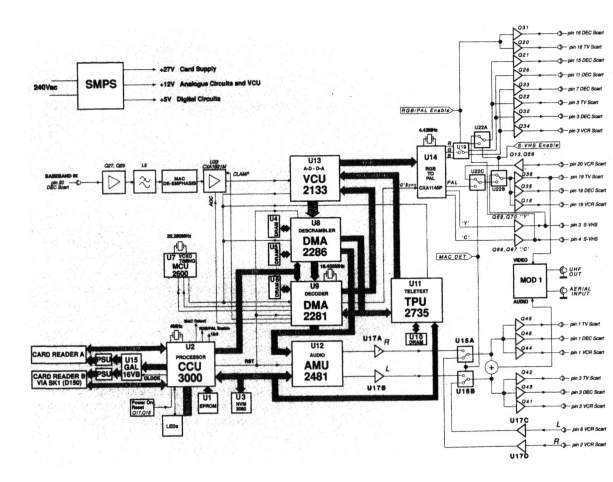

Figure 13.3 *Block diagram of D2-MAC decoder (Pace)*

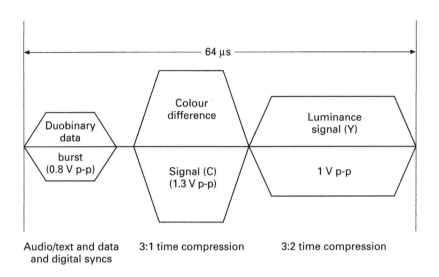

Figure 13.4 *MAC line period*

D2MAC decoders

It should be clarified that MAC (Multiplexed Analogue Components) is a parallel with PAL, SECAM or NTSC as a video format – although it is always broadcast in a scrambled format – whereas if it were encrypted it would be most likely to use the EuroCrypt system which is thus analogous to Videocrypt. Figure 13.3 illustrates a block diagram of a D2-MAC EuroCrypt decoder which can be connected to a PAL receiver via the decoder scart socket. Such a unit with a tuner, etc., could be purchased as a MAC receiver. The basis of operation has much in common with Videocrypt as described but MAC of course has the difference of having luminance (Y) and chrominance (C) separate in its line period (Fig. 13.4). This lends itself then to presenting an output in the Y/C format via modified scart configuration or S Video connector.

The operation of the decoder can be compro-mised in many of the ways already discussed for Videocrypt – signal problems – check PAL tele-text, if corrupt across most channels, then this is likely to indicate that a MAC decoder problem is due to signal problems. When problems are encountered with non-operation, try replacement interconnecting leads – as short as possible, anything over 1 m is asking for trouble. Check the routing and de-emphasis are set correctly. Check the decoder with a different model or brand of receiver – you may have a compatibility problem – specifically the level or condition of the video signal from the receiver used.

Signal tracing around a MAC decoder is much as for Videocrypt. 7-bit digital video during the descrambling process and then analogue in and out. Do not get led astray by faults, especially surrounding conditional access, where unofficial or 'pirate' cards are being used. Suspect these first!

14

DIGITAL TELEVISION AND MPEG-2

At the time of writing, we are just seeing the beginnings of the biggest revolution in television since the introduction of colour transmissions. Digital television systems are now in existence via satellite (see Fig. 14.1) and will soon be available terrestrially in the UK. The principles involved are rather more the province of mathematicians and computer software engineers than consumer electronics engineers, but you won't find either of the aforementioned repairing the reception systems, so major effort is required on the part of ourselves to learn all about it. To date, naturally, there is little servicing experience. Advice can be given on the basis of logic but we shall see the relevance of this as future editions of this book build on experiences. As with the author's previous titles, where we have these circumstances, he tries to provide a service engineer's guide to the technology involved. For a more detailed analysis of digital television systems and a wider coverage of possible reduction and modulation techniques, refer to *Newnes Guide to Satellite TV* by D. J. Stephenson from the same publisher. For all systems, for reasons that will become clear, there is still a major question mark over quality and perception among the viewing public. It is possible to sacrifice quality of picture to give dramatic capacity improvements in a given bandwidth channel. It is likely that initially, many 'faults' will be viewers' objections to the nature of digital transmissions. Audio is largely disregarded in this description but is similarly reduced and then multiplexed into the data-stream in packets.

The most commonly employed format for digital transmission via satellite is MPEG-2. (Motion Pictures Experts Group). This is the form that we shall consider in this chapter. To receive digital signals, even if from a satellite that is providing analogue ones, a completely separate receiver is required. There is no potential for a plug-on decoder. Therefore a simple, mini-distribution system may be the order of the day to feed separate analogue and digital receiver/decoders. In the relevant areas throughout the book we have highlighted areas where particular attention has to be paid when considering digital signals – the cabling, distribution system and LNB. This chapter aims to consider the receiver/decoder end of the operation.

Principles of digital video

Instead of transmitting the analogue video signal, it is analogue-to-digital converted. To provide a sufficiently accurate representation of the analogue video, a suitably high sampling rate must be chosen but the higher the sampling rate, the greater the required bandwidth for transmission. This quandary is the basis of the Nyquist statement which states that the sampling frequency must be at least twice the highest sampled frequency. Having sampled the signal, it is quantised (given a numeric value representing its amplitude) and this value is binary encoded. We thus achieve a stream of binary data. However, to achieve the required accuracy, the data rate required is huge and requires more bandwidth than an equivalent analogue signal.

Redundancy

It can be seen, from a common-sense point of view, that between frames of an analogue video signal, it is likely that much information remains largely unchanged. In a scene of a person walking along a road, the camera is stationary and the person walking through the shot, the road, hedges, sky, etc., will remain largely unchanged from frame to frame, the person will be the main area of moving information. Would it be possible, therefore, to save bandwidth by only transmitting the changes rather than repeating lots of stationary information? The answer is yes and it's called data compression by redundancy – this is called

Digital Television

XP	FREQ.	BEAM	CHANNEL	CODING	ENCRYPTION	HOURS	LANGUAGE
EUTELSAT II-F1 and HOT BIRD 1 (13° East)							
2V	11.238	Superwide	MTV (Italy)	DVB MPEG-2	Cryptoworks	24	English/Italian
	11.238	Superwide	VH-1 Germany	DVB MPEG-2	Cryptoworks	24	German
	11.238	Superwide	Sci-Fi Channel	DVB MPEG-2	Cryptoworks	18	English
	11.238	Superwide	MTV Europe	DVB MPEG-2	Cryptoworks	24	English
	11.238	Superwide	MTV (Germany)	DVB MPEG-2	Cryptoworks	24	English/German
	11.238	Superwide	Bloomberg	DVB MPEG-2	none		English
4V	11.283	Superwide	Telepiù 1	DVB MPEG-2	Irdeto		Italian
	11.283	Superwide	Telepiù 2	DVB MPEG-2	Irdeto		Italian
	11.283	Superwide	Telepiù 3	DVB MPEG-2	Irdeto		Italian
	11.283	Superwide	Discovery Channel (US)	DVB MPEG-2	Irdeto		English
	11.283	Superwide	MTV Europe	DVB MPEG-2	Irdeto	24	English
	11.283	Superwide	CNN International	DVB MPEG-2	none	24	English
33H	11.610	Super	RTL TV	DVB MPEG-2	none	24	German
40H	12.621	Super	AB Channel 1	DVB MPEG-2	none		French
	12.521	Super	AB Encyclopédia	DVB MPEG-2	none		French
	12.521	Super	AB Animaux	DVB MPEG-2	none		French
	12.521	Super	AB Polar	DVB MPEG-2	none		French
	12.521	Super	AB Musique Classique	DVB MPEG-2	none		French
	12.521	Super	AB Cartoons	DVB MPEG-2	none		French
	12.521	Super	AB Charm	DVB MPEG-2	none		French
	12.521	Super	AB RIRES	DVB MPEG-2	none		French
45V	12.542	Super	Telepiù 1 §	DVB MPEG-2	Ideto		Italian
	12.542	Super	Telepiù 2 §	DVB MPEG-2	Irdeto		Italian
	12.542	Super	Telepiù 3 §	DVB MPEG-2	Irdeto		Italian
	12.542	Super	Discovery Channel (US) §	DVB MPEG-2	Irdeto		English
	12.542	Super	MTV Europe §	DVB MPEG-2	Irdeto	24	English
	12.542	Super	CNN International §	DVB MPEG-2	none	24	English
41H	12.567	Wide	USIA Europe	MPEG-2	none		English
	12.576	Wide	Pro TV	MPEG-1.5	authorization only	24	Rumanian
46V	12.583	Wide	TF1	DVB MPEG-2	none	24	French
	12.583	Wide	France 2	DVB MPEG-2	none	24	French
	12583	Wide	Supervision	DVB MPEG-2	none		French
	12.583	Wide	France 3	DVB MPEG-2	none	20	French
	12.583	Wide	La Cinquième	DVB MPEG-2	none	13	French
	12.583	Wide	arte	DVB MPEG-2	none	8	French
	12.583	Wide	France 2 (réseau 2)	DVB MPEG-2	none		French
	12.583	Wide	M6	DVB MPEG-2	none	24	French
EUTELSAT II-F2 (10° East)							
38V	11.631	Wide	Kral TV	DVB MPEG-2	none		Turkish
EUTELSAT II-F3 (16° East)							
20H	11.015	Super	TV10	DVB MPEG-2	none		Dutch
20H	11.024	Super	The Music Factory	DVB MPEG-2	none		Dutch
21H	11.060	Super	Wiszle TV	DVB MPEG-2	none		Polish
40H	12.521	Wide	SiSLink (up to 6 channels)	MPEG-1.5	authorization only		English
41H	12.577	Wide	SNAI – Diretta TV	DVB MPEG-2	none		Italian
46V	12.600	Wide	Number 1 TV	DVB MPEG-2	none		Turkish
EUTELSAT II-F4 (7° East)							
27V	11.135	Wide	Antena 1	DVB MPEG-2	authorization only		Rumanian

Figure 14.1   *List of digital TV services from the Eutelsat craft (Eutelsat)*

temporal redundancy, *Statistical redundancy* encompasses the principle that future frame information is predictable by virtue of being often repetitive. Therefore less data need be transmitted if the known information is used to predict future frames.

Considering the same image, we can see that large parts of the picture will have the same colour and intensity as the adjacent ones – blue of the sky for example. Further savings on data can be made by not transmitting this repeated data – this is *spatial redundancy*. There is also the potential to reduce the amount of information sent by allowing for the fact that the human eye tolerates certain image distortions without the human being aware of their presence. This form of reduction is based on *subjective visual* redundancy.

Source image coding

Figure 14.2 gives a simple overview of how this redundancy is introduced to the signal process. The data-stream of sampled video enters at the left and redundancy techniques are applied. Thus we have a fixed data rate in and a variable one (dependent on the video signal content) out. The buffer stage stores a few frames of data and averages them to produce a fixed data rate again for transmission for MPEG-2. The maximum permissible video data rate is 15 MBit/s. It can be seen then that if a signal has a lot of fast moving action, the amount of redundancy applied will be small. In certain circumstances, the data rate will be too great for the buffer and so the feedback shown will add tailored distortion to reduce the data rate. The data can be reduced to between 1 and 5% of its original content! The earlier point about effects on screen and viewer acceptance is

highly applicable here! One is likely to see image distortions in images such as a racing car driving past a stadium at high speed.

Application of redundancy

This is where the book of hard sums is brought into play! The techniques used thoughout digital reduction vary but the format used in our sphere of interest is *discrete cosine transform* coding. The theories are immensely mathematical but are very interesting should you have a mind for such things. Either way this level of theory is of little relevance to us here, we need consider that the transform is all carried out within an IC which transforms the raw samples of video into a set of coefficients representing the spatial frequencies involved. There is data developed from predicted frame content based on previous content. The transmitted reduced data must obviously be 'rebuilt' at the receiver by a decoder.

The data-stream for MPEG-2 begins with a 1 byte header followed by 187 bytes of 'payload'. The first few bytes of this data are the PID (packet identifier) which advises whether the packet contains audio, video or data, and which programme or service it is part of. Following this part of the stream is a group of 16 bytes of forward error correction code. Thus each packet requires 204 bytes.

Transmission modulation methods

The digital data signals are transmitted using a phase modulation technique. In our case this is

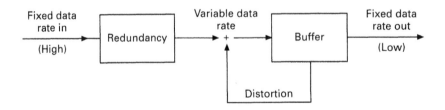

Figure 14.2   *Block diagram of the redundancy introduction process*

likely to be QPSK (Quaternery Phase Shift Keying). Here we have a pair of reference axes, I (in-phase) and Q (quadrature). A change in the bit stream between 1 and 0 gives rise to a phase change of the carrier (Fig. 14.3). This system is energy efficient and robust against interference from other digital signals. The demodulation process is shown in block form in Fig. 14.4. The first i.f. is applied to two product detectors and a reference carrier generator. The carrier is regenerated from the received signal and so gives rise to little in the way of phase error. The two product detectors are 90° phase shifted from each other and by analysing their respective outputs, the correct two bit data (di-bit) can be achieved.

Receivers

Figure 14.6 shows a block diagram of a digital satellite TV receiver. It is quite possible that combined satellite and terrestrial digital receivers will be available shortly. The only different areas are in the tuner, ADC, QPSK demodulator and FEC decoder. We will consider relevant blocks. The tuner works in the same way as an analogue receiver. The carrier is removed to produce I and Q signals but this is under the control of the QPSK demodulator. The ADC converts the analogue I and Q to 6-bit digital outputs components via a pulse width modulator.

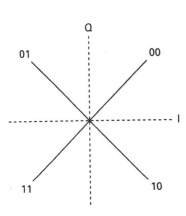

Figure 14.3   *QPSK vectors (phasor diagram)*

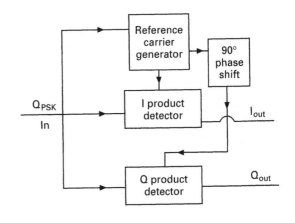

Figure 14.4   *QPSK demodulator principle*

Within the QPSK demodulator, I and Q datastreams are formed into *soft decision* data-streams which are approximations of the data fed into the transmitter modulator. This stage is prior to forward error correction and so the data here is not known to be correct. 3-bit numbers emerge.

The FEC decoder makes *hard decisions* using the redundant FEC information from the datastream and thus taking the 3-bit numbers from the QPSK demodulator, produces MPEG data. The FEC decoder determines whether packets contain errors, corrects them and flags any that it can't correct.

The transport demultiplexer sorts the MPEG data into services, audio, video and data and

There are other phase modulation methods used in digital video. An extension of QPSK is 8PSK where eight vectors as opposed to four are used, similarly 16PSK. By adding the feature of amplitude modulation to the phase vectors, one achieves another used TV transmission system – 16QAM (Quadrature Amplitude Modulation). See Fig. 14.5. Two amplitude levels can be seen on four phases at 90° to each other which are extra to the 8PSK vectors.

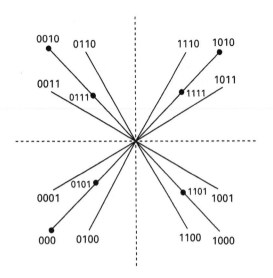

Figure 14.5   *16QAM vectors (Phasor diagram)*

routes them accordingly. This it does by examining the PID. The video information is reconstructed from the compressed data transmitted within the MPEG video block. Parallel data enters and 24-bit video emerges. This video is digital component and so has to be PAL encoded.

Data and communications

The introduction of digital service receivers has brought with it some rather new features. As well as the smart card type of conditional access that we are used to, a CA module can be plugged in to the unit externally. Hence over-the-air conditional access is again with us. A modem may well be incorporated into the receiver for communication with service providers – home shopping being the obvious example for its use. A serial port (RS232) is provided for analysis and upgrading of software. Software can also be upgraded over the

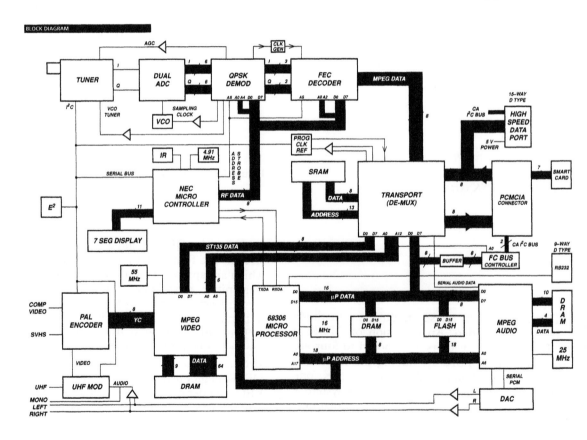

Figure 14.6   *Block diagram of digital satellite receiver for MPEG-2 (Pace)*

air. Within the receiver, we can see buses and data communication rather more at home in a PC, UARTs, etc.

Servicing

As with all new technology, it is very easy to be blinded by it. The standard set of basic fault finding rules apply. First check power supplies for level and cleanliness. Check oscillators and resets. Check status lines. Look at the eye pattern from the demodulator and check for level. Do not overlook the plethora of set-up problem possibilities available via the receiver menus, and don't overlook the possibility of a subscription problem with pay services.

These receivers contain ICS that have over 150 pins and they are very close together. PCBs are multi-layer. Do not, under any circumstances, attempt to replace devices if you are not equipped and capable. Specific advice on these procedures is given in the next chapter.

Most receivers will provide an indication of no signal – or what it perceives as no signal – i.e. no MPEG data. Naturally this will result in no sound or vision but equally all on screen menus etc., should still be working.

The nature of digital receivers means that a message on screen may be the same as when there is signal but it is not tuned, e.g. 'Searching for frequency' or similar. Again, basic checks need to be applied. Is there LNB supply and are there signals entering the receiver? Beyond this one can begin to look at the tuner/QPSK decoder area. The latter can be checked for presence of the aforementioned I and Q signal inputs from the tuner and if no output data is present then again check the vital requirements of supply and clock.

If data is present up to the decoder but one is presented with no sound or vision we need to consider the perhaps worrying prospect of a decoder fault. There are plenty of supply and clock checks to be made again when you have established that good, constant data is arriving from FEC. Unfortunately the close working relationship of the decoder, system control and conditional access sections of these units means that it is very easy to overlook one affecting the other.

One heartening point about the circuits of such highly integrated and complex units as these is that outside the VLSI devices there are few peripheral components. It is almost like fault finding using a block rather than schematic diagram! As stated for bus activity in Chapter 11, it is not possible to analyse data signals but conforming their presence at the correct level is usually sufficient evidence to look elsewhere for your problem.

15

REPAIR TECHNIQUES

In this chapter we shall consider the way to go about actually carrying out repairs once the fault has been established. Most engineers are already proficient at the 'physicals' of repair work, but those that merit special mention are covered below: the first few relate to problems which could well cause the sorts of intermittent faults which were discussed in the introduction.

Printed circuit boards

PC boards are a feature of virtually all consumer electronic equipment, and are a prolific cause of trouble, generally with open-circuit paths of one sort or another.

Dry joints

One of the most common repairs required is that of resoldering a poor joint. The term 'dry joint' is used as a blanket description for this type of fault but the problem could equally be due to corrosion. The best policy with all such repairs is to first remove all the existing solder then clean the component leg or lead and tin it before resoldering. If the print land is also damaged, expose some more print by scraping off the lacquer and solder the leg or lead onto this new area: the safety aspect should always be paramount when doing this type of repair. Ensure that the joint you make is mechanically and electrically good: if the component leg or lead has become damaged (for example due to arcing in the dry joint) replace it; where high or low melting-point solder was used in production, use the same type during repair.

With modern production techniques dry joints can occur anywhere in circuit but they are most likely on PC board-mounted controls and switches; heavy components; thick- or very thin-legged components and parts of the circuit which carry heavy currents or run hot. An increasingly common problem that can be extremely difficult to detect is where there is apparently a dry joint, but in fact the leg of the component has not been pushed through the hole in the PC board prior to soldering in production and so makes intermittent contact with the cap of solder over the hole.

Provoking a dry joint is generally done by tapping, probing or flexing the suspect area, once the all-important visual inspection has proved fruitless. If the effect of a dry joint is not destructive or dangerous, and the dry joint found is not conclusive proof of having cured the fault, disconnect the joint fully and see if you get the same symptoms. As with all intermittent problems, once you get the fault to occur don't try to get rid of it – use traditional (but gentle) fault-tracing methods to find where the break, 'dry' or failure, is.

It may seem tempting to blanket resolder an area where a dry joint appears to be present. This can be somewhat dangerous as it can clear faults temporarily, thus masking the fact that you haven't actually traced the problem. For example, you may heat up a crack in a PC board which, when cool again, will re-open, or have heated up an electrolytic capacitor that is only faulty when cold. It is far better to cure the fault first then re-solder anything suspect. When blanket re-soldering do take care not to introduce shorts – it is easy to do when you see lots of poor joints and you work rather too quickly.

Print breaks

These are the (usually invisible) defects that can lead to a lengthy investigation of a nasty fault. Visual checks on top and bottom of the board are recommended here as a stress mark can often be seen on the component side. Otherwise, locating a print break involves flexing the PC board – one way opens the break and prompts the fault; the other closes the break and clears the fault. This could also be due to a dry joint but that is less likely. The way in which the fault reacts depends on which side of the board the flexing takes place – and in the case of double-sided print, which side the break is on. Voltage checks often reveal a break, where there are no intervening components, but do watch out for ICP type fusible devices, which manufacturers tend to insert in production but not necessarily show on the circuit diagram. When looking for any fault it is wise to scan over all the components fitted to the print side of the PC board(s) to ensure that none have broken off or caused a short-circuit.

Continuity checkers/bleepers as found on many modern meters can be useful, but do not fall into the trap of taking a 'beep' to indicate a true short circuit: most meters beep up to 200 Ω or so, and so you could be reading through a coil or transformer winding. The fact that any component check is best made out of circuit bears mentioning here – problems such as reading through parallel paths and the meter being affected by charges stored in capacitors or batteries can cause you to make a wrong diagnosis.

Breaks in print tracks should be repaired with wire run from the nearest lands on either side of it. Breaks in 'open' (non-lacquered) carbon print can be repaired with conductive paint, but beware of introducing short-circuits to adjacent print! Too much paint could render the track too thick – this can be a problem in remote handsets where, if the print is too thick, the contact could be closed.

It is wise to drill a small hole at either end of a crack (if it is in the board as well as in the print) to prevent it speading further.

It is common these days to find PC boards where print is layered, i.e. there are different tracks on top of each other and obviously insulated. Unfortunately, this introduces a symptom that can be horrendously difficult to find – electrical leakage between the two layers. Isolate both ends of each track and measure them to try to prove this, but linking out the area can be the only answer.

Many faults can be caused by glue applied to PCBs – electrical leakage results as the glue ages – remove it all thoroughly to prove.

Board damage

Where the print is damaged there may be damage to the PC board itself. If a board is burnt a small amount can be cut out, again allowing for the safety factor, and the conductors hard-wired. If more than a few tracks need linking or if the PC board is holed, especially in power supply areas, then it would contravene BEAB approval (in the UK) to try to repair it. The only safe answer is to replace the PC board. Do consult with the manufacturer in such circumstances – there have been instances where certain designs have been prone to PC board damage due to a weakness in soldering, for example, and manufacturers have special schemes for PC board replacement or, indeed, a modification to do away with the affected area of the PC board. If a board is cracked (and space permitting) a new sheet of paxolin could be glued over the surface to provide strength or the hole filled with epoxy resin. Small cracks can be stabilised by repairing print breaks with stiff wire which also serves to support the board. It is unwise to hard-wire power circuits.

Liquid spillage

The entry of liquid to any unit can wreak havoc. Caught early enough it can sometimes be cured if confined to a small area of PC board. This is seldom the case in practice because faults caused by spillage tend to occur only after the liquid has lain inside for a while, when corrosion sets in: the majority of customers deny all knowledge of the accident, which makes life difficult if the substance cannot be identified. Excepting minor cases, caught early, the best policy with liquid spillage is to replace all affected parts: if this means that the repair cost is prohibitive, so it must be. Experience shows that future reliability cannot be guaranteed and a repair is seldom fully effective.

In minor cases it is best to replace all parts affected to give the best chance of a future reliability: particular attention should be paid to mechanical components. It is necessary to clean under components by removing them first. This includes SMDs (surface mounted devices).

Before giving the almost obligatory estimate for this type of repair, make sure that the parts required are available: components which rarely fail electrically or mechanically (but may well be wrecked by corrosion!) may not be stocked by the manufacturer.

Surface mounted devices

You cannot fail to notice SMDs these days. To most the reasons for their use are obvious. They are in themselves very much smaller than their conventional through the board (TTB) counterparts, and because they mount on the surface of the board then can be used to form two entirely separate circuits on either side of a PC board. Production techniques are simplified as components are glued in place and then wave-soldered so reliability tends to be higher – dry joints are much less likely.

Problems do occur at the service end with all this new technology, however: one major problem is the miniscule size of the devices. Fig. 15.1 shows some examples of SMD type ICs for those not familiar with them. Maybe at present when a faulty SMD requires replacement it is ordered direct from the manufacturer concerned and thus you don't really need to know how to decode the values – especially when you have the circuit diagram to guide you. In the future, though (it is already the case in many workshops) they will be held in stock in the same way as conventional components: here it's essential to be able to determine the value or type.

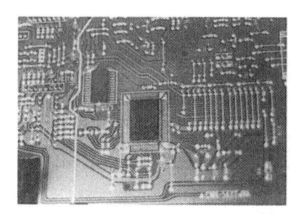

Figure 15.1 *A SMD flat-pack IC, a common sight in modern electronic equipment. Replacing and handling calls for specialist techniques, as discussed in the text*

Apart from the size problem we also have to deal with actually fixing them to the board as well as soldering them. On top of this most SMD semiconductors are static sensitive, calling for special anti-static precautions in storage and handling. The ICs, while shrinking in physical size, have ever-growing numbers of lead-outs. It is common to find ICs with them on all four sides, sometimes totalling over 200: traditional soldering techniques need considerable adaptation to deal with SM assemblies.

Desoldering

Let's first discuss the removal of SMDs, using our existing, traditional (25 W or so) bench soldering iron.

For small two- or three-connection components the method is fairly straightforward: a technique similar to that used for traditional style components can be used. Simply remove as much solder as possible from the joints, then hook each leg up and off with either the tip of the iron or a probe such as a flattened-out staple or paper clip. Many SMDs are glued to the PC board: application of slight heat melts this sufficiently to remove the device (usually!). This is best done in a flick or twist action. The most effective remover of solder is desoldering braid, the finer the better: it is widely available in 0.5 mm width from spares wholesalers. The main thing to remember is to use the minimum amount of heat possible, in practice with a low-power iron and the shortest dwell-time of iron and braid on the joint. Heat damage to the component is not the prime consideration when removing a component which is faulty anyway (although modern semiconductors will withstand surprisingly high temperatures). The print on modern PC boards is usually very fragile and it lifts or the pads for the SMDs break off print tracks if too much heat is applied – and too much needn't be very much! Obviously, one also tries not to damage the component when removing it if it is only being taken off for accurate testing or stage isolation/load lifting. If the smaller two- or three-connection device uses contact pads rather than legs, once the solder is removed the component can be twisted off with a pair of tweezers – and with gentle heat applied if glue is holding the body.

For multi-connection devices such as ICs there is a choice. The first method described below is ideal for smaller ICs with up to, say, 28 legs.

Figure 15.2   *Hot-air 'Pyropen', a remarkably useful device which is ideal for small-scale re-working of surface mount devices. It can also be used as a soldering iron, as in this illustration. The cost is around £50 (UK). It is gas powered (Weller/Cooper Tools)*

Many consider the same method preferable for all ICs, but the second method is an alternative.

Method one

Employing the same technique and precautions as above, remove as much solder as possible from the connections, then lift each leg in turn with the probe. An alternative approach to this is to solder a piece of thin wire to an adjacent land, feed it through all the pins of one side to be desoldered and as each leg is heated pull the wire's end to progressively release them. This is known as the 'cheesewire method'. With all legs pointing skyward, gentle heat at either end of the body of the device is sufficient to melt the glue invariably used under devices of this size. Again very carefully clean up excess solder from the pads on the PC board and remove dirt and flux from the area with isopropyl alcohol or a similar solvent.

Method two

The alternative method for removing larger devices is to cut the legs of the device off, remove the body and then unsolder each leg in turn with a 'flick' action of the soldering iron. This reduces the likelihood of heat damage to the print but naturally renders the device useless. It is also very easy to damage the print. Use multiple, gentle knife strokes.

With the legs removed the PC board should be cleaned up and prepared as for method one.

Method three

The most elaborate method of removing an SMD, particularly applicable to multi-legged devices, is the hot air *rework* type iron, where a hot air jet is used instead of a bit to melt the solder. Commonly incorporated in such devices is a vacuum pump to remove the molten solder. As with more elaborate irons, temperature control is provided and should be used with all modern delicate PC boards.

There are now various hot-air tools that do not remove the solder, but do allow all pins to be heated simultaneously to allow the device to be flicked off. Some tools provide a single jet of air, e.g. the Weller Pyropen in Fig. 15.2. This is a very reasonably priced tool (around £50–60 in the UK) and is extremely useful. More expensive units with fan assistance also use multinozzle fixtures that match the IC package exactly, making the task very simple. The disadvantage is that you need one of each nozzle for all the IC packages you encounter, and they are expensive. Furthermore, the heat is not applied directly to the body of the device and so any glue under it may well not be melted and thus the device will not move, even when all legs are heated. Some manufacturers' glues are rather stubborn even when directly heated. If you encounter this situation, do not continue heating the device – use method one and then twist the device off the board.

A relatively new version of this type of equipment is now available where the heat is generated by a u.h.f. transmitter delivering its output into a tuned load which is the appropriate tip of the iron. This has a number of advantages with regard to speed of heating and thermal regulation, and costs seem to be very competitive.

Method four

Attachments are available for some soldering irons, shaped to cover all the pins of an IC at once in order to be able to 'heat and grab' the device. They can also be used on through-the-board devices. Their use depends on availability and the overriding factor must be the amount of heat applied during the process.

Resoldering

As with desoldering and removing the device there is a choice of methods of fitting the replacement.

Method one

Prepare the connection pads on the PC board by tinning them with low melting point (LMP) solder. Do not use too much solder otherwise the device, particularly if it is a multi-legged one, will tend to 'roll-off' the pads. With the PC board prepared, offer up the device and if you want to glue it in place do so. Most engineers see no need to do so in a 'rework' situation. Glued or not, the component must next be soldered in. Simply touch each pad with the iron while introducing sufficient extra solder to make a good joint. Solder corner pins first ensuring that the device is sitting square before doing the rest.

Method two

This method is suitable for multi-legged components such as flat-pack ICs, but is rather unnecessary for two- or three-connection SMDs.

After removal of the faulty device and board cleaning the solder pads are not tinned with solder – instead solder cream is spread in a line over the row(s) of pads. This is a paste containing 'solder-dust' suspended in flux; the solder has an extra (about 2%) silver content to allow it to flow more readily. When heated the paste amalgamates on the pads leaving a very quick, neat solder job.

Figure 15.3 *The standard static-sensitive warning symbol*

Gauging the correct amount of solder cream to use is crucial: too much leads to inter-pin shorts; too little leads to dry joints. Place and fix the device in the same way as for method one and then simply work around its pins. If the device is an IC with pins on all four sides it is best to fix one pin on each side at a time, starting in opposite corners: this ensures that the device is square before too much work is done. It is upsetting to say the least to find that it is off-centre when starting to solder to final row of 20 pins and having to unsolder the 60 already done! Finally clean up excess residue with alcohol. This method is best facilitated by use of a hot-air tool, as described earlier.

Anti-static precautions

Another problem that today's engineer has to cope with is the ever-present danger to many components from static charges, present everywhere. The majority of modern semiconductors are static-sensitive. Fig. 15.3 illustrates the warning symbol used to denote static-sensitive devices. The important thing to remember is that while static damage can destroy a device immediately it very often results in a shortened life: thus the damage may not immediately be apparent.

Static-sensitive devices should be marked as such, and supplied and stored in protective packs, usually consisting of conductive plastic bags, or boxes lined with conductive foam. There are now available anti-static storage drawers and racks which enable the best protection against electro-static discharge (ESD) when storing components. When considering ESDs, if in doubt always take precautions – sound advice!

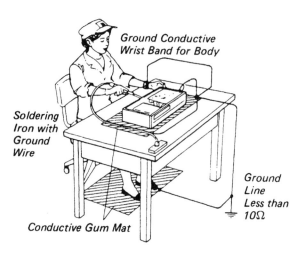

Figure 15.4 *A suitable bench set-up for anti-static working*

Protection

The way to guard against static discharge is to ensure that your body, the connections to the device, the body of the soldering iron and the ground of the appliance are all at the same potential. Of course equipment under repair should not be powered during such activities! Wrist straps can be used to provide a ground connection, but the most successful and practical approach is to provide a small area of bench specifically for replacing static-sensitive devices, equipped with a conductive mat surface and floorcovering. Portable mats are available as well as 'build-in' kits: see Fig. 15.4.

Naturally such an earthed area cannot be present in an isolated area of a workshop (used for servicing live chassis equipment) and so distance and physical barriers should prevent contact between the two.

Some commonsense precautions also minimise the risk of static damage: keep devices in their protective packing until the last moment, and try to keep all lead-outs shorted together until the device is soldered in.

It is always worth taking similar precautions when removing a suspect component – it may not be faulty after all! With this in mind, ensure that any desolder tool used is an anti-static type.

Use of variac, bulbs and dummy loads

Great advantage can be gained, when repairing a fault, in supplying an externally-generated operating voltage to part of a circuit with the normal supply disconnected; and by varying this supply, starting at a low level and gradually increasing it to the normal operating level.

The most practical way of providing a variable supply is with the *variac*, a mains transformer with variable output from OV to full mains potential. The unit under test is fed from it, and in cases where a 'soft start' or low voltage supply are required (either during fault finding or following a tentative repair) it is ideal: such situations arise when repairing a power supply or amplifier output stage. Where the fault causes immediate fuse blowing at power-up, running with a low supply voltage allows electrical and physical checks to be made with a low, safe current flowing. This and powering units that require a low a.c. supply.

Bulbs or dummy-load resistors can also be very useful and are often used in conjunction with a variac or external supply. Where it is suspected that a supply line is being 'pulled down' by an excessive load, a substitute ('dummy') load can be fitted across the rail and the voltage rechecked to confirm that it is not (e.g.) just a case of poor regulation.

In certain circumstances, and when appropriate checks have been made it is possible to fit a bulb in place of a fuse: for example, to give a visible indication of a protection circuit operating and to allow, for a period, checks to be made on a circuit which would otherwise be shut down by the fuse blowing. Of course one has to be careful as to when this technique is used: nasty damage to the circuit and yourself could result. The most fruitful application is when an intermittent fault is suspected to stem from an overload in a particular area.

Thermal faults

There are two approaches to tracing a *thermal fault*, one which comes or goes when the unit is either cold (i.e. at switch on) or hot (after a period of running). The approach depends on which of the two instances applies. The suspect area or component can be heated or cooled to provoke or clear the fault as the case may be. In the author's experience it is always best to try to clear the fault rather than provoke it because extremes of temperature can fall outside the tolerances of the components and thereby *cause* faults. Therefore if a fault is cleared by application of heat or cooling it is far more conclusive than a fault that is instigated thus.

To heat components the tip of a soldering iron or hairdrier can be used, but in the former case care should be taken not to apply a mains-powered or earthed iron in an inappropriate place! To cool components chemical aerosol freezers are ideal. When heating or cooling a suspect area or component try to narrow the application as much as possible by using nozzle-extension tubes on aerosols and barriers around the area to prevent the effect spreading to other components. These can take as simple a form as a piece of cardboard.

Some precautions need to be borne in mind when dealing with apparently thermal faults. Firstly, the problem could well be due to a break in the print or a dry joint: heating or freezing may not show this up as successfully as when a faulty component is present. Secondly, component characteristics drift with temperature, so a component can be perfectly OK but because of its change in characteristic with temperature a fault symptom develops. The same effect can arise from other factors, like an adjustment set at the edge of its range: an example of this is incorrect setting of the quiescent current in an audio output stage leading to sound distortion when the output stage is hot. Take care, then, in applying artificial heat and cooling; experience counts for a lot here.

Appendix 1

REFERENCE DATA

There is a certain amount of information that is general to satellite TV servicing that cannot be pigeon-holed into one of the rather specific contexts of the chapters of this book. Similarly other information is not specific to satellite – general consumer electronics data that is necessary to enable us to interface satellite systems with the rest of the world. Herewith then a compendium of these matters.

Resistor colour code

The colour code is no longer the only way of marking resistor values as it was on lower wattage types. The advent of the SMD brought the need for a different method, explained in Chapter 15. This excepts the MELF (metal ended leadless frame) type of device which still uses the colours.

The system works by marking four coloured bands on the body of the resistor: the first two are the digits, the third the multiplier and the fourth the tolerance value. As can be seen from Table A1.1 it is easy to establish the direction in which the bands should be read: the tolerance and multiplier bands are restricted in the number of colours they use.

If we take as an example, RED, PURPLE, YELLOW, GOLD, this means that the digits are 2 and 7 and the multiplier, yellow, indicates a factor of ×10 000 – or add 4 noughts, making the value 270 kΩ with a tolerance (gold) of 5%: it should, then, be within the range 243 to 297 kΩ. If, of course, this range is unacceptable for the operation of the circuit a closer tolerance resistor would have been specified by the designer.

Table A1.1 also shows the coding for temperature coefficient (a fifth band), not always used in practice.

Surface-mount components use a different method of value identification, excepting certain MELF type devices which use the standard colour code. Transistors and diodes usually use a two character code which has to be checked with a conversion chart in the appropriate service manual. Resistors and capacitors generally have a three-digit code: the first two digits give the value and the third the multiplier. For example, '103' on a resistor would equate to 10 kΩ – 10 plus three zeros = 10 000. Capacitors often have no markings on their traditionally brown bodies. Reference must be made to service information. Usually there are no circuit references printed on PC boards for SMDs. Use of a print/component layout is necessary.

Decibel scales

The decibel, symbol dB, is one-tenth of a *bel* which expresses a ratio between two levels. The scale is logarithmic to comply with the human aural perception ability. Table A1.2 shows a decibel scale of common values against voltage, current and power ratios.

As an example we can say that an amplifier with a gain of 20 dB and an input signal of 100 mV p–p will have an output of 1 V p–p and a gain of 10.

Plugs and sockets

Figure A1.1 illustrates some common plug and socket connections that may be encountered.

The scart connector

This connector (aka Peritel or Euroconnector) is now the standard analogue A/V interface. (In the forthcoming digital equipment, the equivalent will be the Firewire connector.) It allows stereo audio in and out, composite video in and out, RGB and blanking, data and status switching (via pin 8)

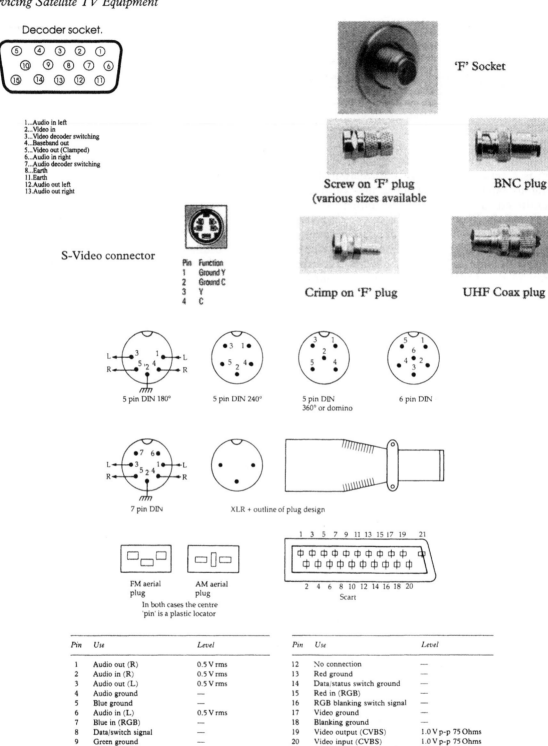

Decoder socket.

1...Audio in left
2...Video in
3...Video decoder switching
4...Baseband out
5...Video out (Clamped)
6...Audio in right
7...Audio decoder switching
8...Earth
11.Earth
12.Audio out left
13.Audio out right

'F' Socket

Screw on 'F' plug
(various sizes available

BNC plug

S-Video connector

Pin	Function
1	Ground Y
2	Ground C
3	Y
4	C

Crimp on 'F' plug

UHF Coax plug

5 pin DIN 180°

5 pin DIN 240°

5 pin DIN
360° or domino

6 pin DIN

7 pin DIN

XLR + outline of plug design

FM aerial
plug

AM aerial
plug

In both cases the centre
'pin' is a plastic locator

Scart

Pin	Use	Level		Pin	Use	Level
1	Audio out (R)	0.5 V rms		12	No connection	—
2	Audio in (R)	0.5 V rms		13	Red ground	—
3	Audio out (L)	0.5 V rms		14	Data/status switch ground	—
4	Audio ground	—		15	Red in (RGB)	—
5	Blue ground	—		16	RGB blanking switch signal	—
6	Audio in (L)	0.5 V rms		17	Video ground	—
7	Blue in (RGB)	—		18	Blanking ground	—
8	Data/switch signal	—		19	Video output (CVBS)	1.0 V p-p 75 Ohms
9	Green ground	—		20	Video input (CVBS)	1.0 V p-p 75 Ohms
10	No connection	—		21	Screening plate surround	—
11	Green in (RGB)	—				

Figure A1.1   *Common plug and socket connections*

which includes automatically switching the input over to scart when pin 8 goes from 0 to 12 V. A third state can be implemented to denote 16:9 wide screen format as opposed to the conventional 4:3. There is a modified configuration to that detailed in Fig. A1.1, to allow S-Video interfacing (as opposed to using the S-Video connector shown). Here pins 19 and 20 become Y (luminance) in and out, and pin become C (Chroma). We have seen in earlier chapters that when used for decoder interconnects, other pin uses are introduced so always refer to manufacturers' literature before assuming a fault – take careful note of what the scart is labelled!

TV systems

As well as considering the transmission standards used by the satellite broadcasters, we have to ensure that the satellite receiver will be providing the correct output in terms of colour system and r.f. frequencies (u.h.f. or v.h.f. and correct sound intercarrier). It is quite possible that units will be received as 'faulty' when actually they have been imported from another market or indeed sent for standards conversion. Table A1.2 lists most standards information that will be required.

Table A1.1 *Resistor colour codes*

Colour	Value	Multiplier	Tolerance	Temperature coefficient
BLACK	0	× 1	n/a	n/a
BROWN	1	× 10	1%	100 ppm
RED	2	× 100	2%	50 ppm
ORANGE	3	× 1000 (1 K)	n/a	15 ppm
YELLOW	4	× 10 000 (10 K)	n/a	25 ppm
GREEN	5	× 100 000 (100 K)	0.5%	n/a
BLUE	6	× 1 000 000 (1 M)	0.25%	n/a
VIOLET	7	× 10 000 000 (10 M)	0.1%	n/a
GREY	8	n/a	n/a	n/a
WHITE	9	n/a	n/a	n/a
GOLD	n/a	× 0.1	5%	n/a
SILVER	n/a	× 0.01	10%	n/a

148

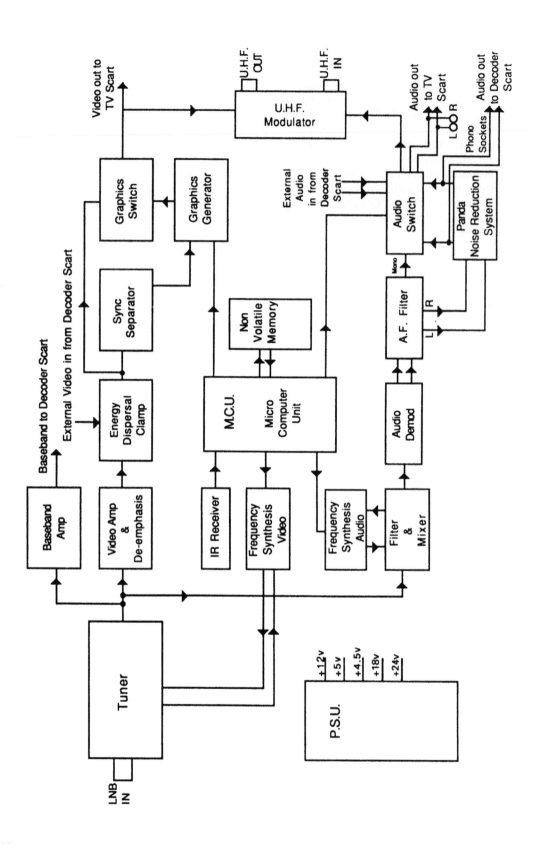

Figure A1.2   Block diagram of complete receiver (Pace)

System	Lines	Channel bandwidth	Vision bandwidth	Sound spacing	vision mod.	Sound mod.
B	625	7MHz	5MHz	+5.5MHz	– ve	f.m.
D	625	8MHz	6MHz	+6.5MHz	– ve	f.m.
G, H*	6.5	8MHz	5MHz	+5.5MHz	– ve	f.m.
I*	625	8MHz	5.5MHz	+6mHz	– ve	f.m.
K, K1*	625	8MHz	6MHz	+6.5MHz	– ve	a.m.
L*	625	8MHz	6MHz	+6.5MHz	+ ve	a.m.
M	525	6MHz	4.2MHz	+4.5MHz	– ve	f.m.
N	625	6MHz	4.2MHz	+4.5MHz	– ve	f.m.

The asterisk indicates that the vestigal vision sideband extends to 1.25MHz; with all other systems vestigal sideband has a bandwidth of 0.75MHz.

The soundspacing is the frequency of the main sound carrier with respect to the vision carrier. Many countries now have stereo/dual sound systems. There are several different systems, as follows:

(1) Zweiton standard B/G. Second sound carrier is at +5.742 MHz. Main carrier is modulated with (L + R)/2 or A, the second carrier with R or B.

(2) Zweiton standard D. Second sound carrier is at +6.742MHz. The carriers are modulated as above.

(3) Zweiton standard M. Second sound carrier is as +4.72MHz. Main carrier is modulated with L + R or A, the second carrier with L – R or B.

(4) Nicam standard I. Second sound carrier is at +6.552MHz. Main carrier is modulated with mono sound or A. The second carrier is digitally modulated with L and R, or A and B, or mono plus data or full data.

(5) Nicam standard B/G. Second sound carrier is at +5.85MHz. Main carrier is modulated with mono sound or A. The second carrier is digitally modulated as (4) above.

(6) F.m./f.m. standard M. Uses a single carrier modulated with an f.m./f.m. multiplex.

(7) MTS standard M. Single sound carrier is modulated with an f.m./a.m. multiplex (supplementary audio programme).

The above system numbers 1–7 are not official: they have been inserted simply to enable us to indicate the system in use in the following World TV Standards List.

Country	System	Colour	Stereo
A			
Abu Dhabi	B, G	PAL	–
Afghanistan	B, D	PAL and SECAM	
Albania	B	PAL	–
Algeria	B	PAL	–
Andorra	B	PAL	–
Angola	I	PAL	–
Antigua and Barbuda	M	NTSC	–
Antiles	M	NTSC	–
Argentina	N	PAL	–
Armenia	D, K	SECAM	–
Ascension Is.	I	PAL	–
Australia	B	PAL	1
Austria	B, G	PAL	1
Azerbaijan	D, K	SECAM	–
Azores	B	PAL	–
US Forces	M	NTSC	–

Country	System	Colour	Stereo
B			
Bahamas	M	NTSC	–
Bahrain	B	PAL	–
Bangladesh	B	PAL	–
Barbados	N	NTSC	–
Belarus	D, K	SECAM	–
Belgium	B, H	PAL	5
Belize	M	NTSC	–
Benin	K	SECAM	–
Bermuda	M	NTSC	–
Bhutan	B	PAL	–
Bolivia	N	NTSC	–
Bosnia	B, H	PAL	–
Botswana	I	PAL	–
Brazil	M	PAL	7
Brunei	B	PAL	–
Bulgaria	D	SECAM	–
Burma	M	NTSC	–
Burundi	K1	SECAM	–

149

C

Cameroon	B	PAL	–
Canada	M	NTSC	7
Canary Is.	B	PAL	5
Central African Rep.	K1	SECAM	–
Chad	K1	SECAM	–
Chile	M	NTSC	–
China§	D	PAL	2
Colombia	M	NTSC	–
Congo	D	SECAM	–
Costa Rica	M	NTSC	–
Croatia	B, H	PAL	–
Cuba	M	NTSC	–
Curacao	M	NTSC	–
Cyprus	B, G	SECAM	–
Turkish sector	B, G	PAL	
Czech Republic	D, K	SECAM	–

D

Dahomey	K1	SECAM	–
Denmark	B, G	PAL	5
Diego Garcia	M	NTSC	–
Djibuti	K1	SECAM	–
Dominica	M	NTSC	–
Dominican Rep.	M	NTSC	–
Dubai	B, G	PAL	–

E

Ecuador	M	NTSC	–
Egypt	B	SECAM	–
El Salvador	M	NTSC	–
Equatorial Guinea	B	PAL	–
Estonia	D, K	SECAM	–
Ethiopia	B	PAL	–

F

Falkland Is.	I	PAL	–
Fernando Po	B	PAL	–
Fiji	B	PAL	–
Finland	B	PAL	5
France	L	SECAM	–
French Polynesia	K1	SECAM	–

G

Gabon	K1	SECAM	–
Georgia	D, K	SECAM	–
Germany	B, G	PAL	1
US Forces	M	NTSC	–
Ghana	B	PAL	–
Gibraltar	B	PAL	–
Greece	B	SECAM	–
Greenland	B	PAL	–
US Forces	M	NTSC	–
Guadeloupe	K1	SECAM	–
Guam	M	NTSC	–
Guatamala	M	NTSC	–
Guinea (Bissau)	I	–	–

Guinea (Rep.)	K1	SECAM	–
Guyana (French)	K1	SECAM	–
Guyana (Rep.)	B, G	SECAM	–

H

Haiti	M	NTSC	–
Hawaii	M	NTSC	7
Honduras	M	NTSC	–
Hong Kong	I	PAL	4
Hungary	D, K	SECAM	–

I

Iceland	B, G	PAL	–
India	B	PAL	–
Indonesia	B	PAL	–
Iran	B, G	SECAM	–
Iraq	B	SECAM	–
Ireland	I	PAL	–
Israel	B, G	PAL	–
Italy	B, G	PAL	1
Ivory Coast	K1	SECAM	–

J

Jamaica	M	NTSC	–
Japan	M	NTSC	6
Jordan	B, G	PAL	–

K

Kampuchea	M	–	–
Kazakhstan	D, K	SECAM	–
Kenya	B, G	PAL	–
Korea N	D	SECAM	–
Korea S	M	NTSC	3
Kuwait	B, G	PAL	–
Kyrgyzstan	D, K	SECAM	–

L

Latvia	D, K	SECAM	–
Lebanon	B	SECAM	–
Leeward Is.	M	NTSC	–
Lesotho	I	PAL	–
Liberia	B, H	PAL	–
Libya	B, G	SECAM	–
Lithuania	D, K	SECAM	–
Liechtenstein	B, G	PAL	–
Luxembourg	B, G, L	PAL and SECAM	–

M

Macao	I	PAL	–
Macedonia	B, H	PAL	–
Madagascar	K1	SECAM	–
Madeira	B	PAL	–
Malawi	B, G	PAL	–
Malaysia	B, G	PAL	1
Maldives	B	PAL	–
Mali	K1	SECAM	–
Malta	B, G	PAL	

Martinique	K1	SECAM	–
Mauritania	B	SECAM	–
Mauritius	B	SECAM	–
Mexico	M	NTSC	7
Micronesia	M	NTSC	–
Midway Is.	M	NTSC	–
Moldova	D, K	SECAM	–
Monaco	G, L	PAL and SECAM	–
Mongolia	D	SECAM	–
Montenegro	B, H	PAL	–
Montserrat	M	NTSC	
Morocco	B, H	SECAM	–
Mozambique	B	PAL	–

N

Namibia	I	PAL	–
Nepal	B	PAL	–
Netherlands	B, G	PAL	1
New Caledonia	K1	SECAM	–
New Zealand	B	PAL	5
Nicaragua	M	NTSC	–
Niger	K1	PAL	–
Nigeria	B	PAL	–
Norway	B, G	PAL	5

O

Okinawa	M	NTSC	6
Oman	B, G	PAL	–

P

Pakistan	B	PAL	–
Panama	M	NTSC	–
Papua-New Guinea	B, G	PAL	–
Paraguay	N	PAL	–
Peru	M	NTSC	–
Philippines	M	NTSC	–
Poland	D, K	SECAM and PAL	–
Portugal	B, G	PAL	–
Puerto Rica	M	NTSC	–

Q

Qatar	B	PAL	–

R

Reunion	K	SECAM	–
Romania	D, K	PAL	–
Russia	D,K	SECAM	–
Rwanda	K	SECAM	–

S

Saba and Sarawak	B	PAL	2
Samoa (Eastern)	M	NTSC	–
San Marino	B, G	PAL	1
Saudi Arabia	B, G	SECAM	–
Senegal	K1	SECAM	–

Serbia	B, H	PAL	–
Seychelles	B	PAL	–
Sierra Leone	B, G	PAL	–
Singapore	B, G	PAL	5
Slovakia	D, K	SECAM	–
Slovenia	B, H	PAL	–
Society Is.	K1	SECAM	–
Somalia	B, G	PAL	–
South Africa	I	PAL	–
Spain	B, G	PAL	5
Sri Lanka	B, G	PAL	–
Sudan	B	PAL	–
Surinam	M	NTSC	–
Swaziland	B, G	PAL	–
Sweden	B, G	PAL	5
Switzerland	B, G	PAL	1
Syria	B, H	SECAM	–

T

Tahiti	K1	SECAM	–
Taiwan	M	NTSC	7
Tajikistan	D, K	SECAM	–
Tanzania	B	PAL	–
Thailand	B, G	PAL	–
Tibet	D	PAL	–
Togo	K1	SECAM	–
Trinidad and Tobago	M	NTSC	–
Tunisia	B	SECAM	–
Turkey	B, G	PAL	–
Turkmenistan	D, K	SECAM	–

U

Uganda	B, G	PAL	–
UK	I	PAL	4
Ukraine	D, K	SECAM	–
United Arab Em.	B, G	PAL	–
Upper Volta	K1	SECAM	–
Uruguay	N	PAL	–
USA	M	NTSC	7
Uzbekistan	D, K	SECAM	–

V

Vatican	B, G	PAL	1
Venezuela	M	NTSC	–
Vietnam	D, M	NTSC	–
Virgin Is.	M	NTSC	–

Y

Yemen	B	PAL	–

Z

Zaire	K	SECAM	–
Zambia	B, G	PAL	–
Zanzibar	I	PAL	–
Zimbabwe	B, G	PAL	

Appendix 2

SAFETY, BEAB AND BS 415: 1990

Throughout this book safety precautions have been described before getting involved in the service process and collectively these form the overall safety instructions here. Some general points also apply. For example, all power plugs, leads and connectors should be checked for correct and safe wiring, correct fuse values, etc. prior to commencing the repair. Always fit the correct type of fuse or fusible device, e.g. quick-blow for quick-blow, not anti-surge. All parts classed as critical safety components should be replaced with exact types approved by the manufacturer of the equipment; these components are often shaded in the circuit diagram and parts list, and are always marked there with either the ⚠ or ⊕ symbols. Cable ties and apparently insignificant blobs of glue or sealant should always be replaced, as should insulating boots, covers, etc., and warning labels. Cabinets and covers should be checked for damage in the form of cracks or holes, and appropriate action taken if any such defects are found. All screws and fixings should be replaced in the correct positions: screws that are too long or short can be lethal.

In the UK there is legislation governing electrical safety testing. Anyone carrying out repair or maintenance to an electrical appliance should test it for electrical safety and record the results.

Appliance testing

The insulation resistance between live/neutral and any exposed metal part should be checked, as should the earth continuity, where applicable (Class II appliances).

There is no requirement for flash testing appliances, and it is advised against due to the damage that can result from incorrect testing methods or too frequent testing.

Portable appliance testers

To facilitate safe, correct and efficient portable appliance testing (PAT), a number of manufacturers in the electrical test equipment business have developed units that carry out all the required tests in one piece of test equipment, into which the appliance is plugged via its mains plug. An external earth connection is made for earth return, where required.

Some very basic units incorporate simply PASS or FAIL indicators, but this does not enable you to record actual values for earth resistance or insulation resistance. The relevance of this is that gradual insulation breakdown, due maybe to a humid atmosphere, can be detected and monitored, as can an increasing earth resistance due possibly to corrosion in case parts.

More sophisticated PATs have a meter with values shown as well as the relevant pass bands. Such a unit is the MEGGER PAT3. This does not have a flash test facility but is ideal for the average brown goods workshop.

More advanced models, which include flash testing, may also enable customising of test routines to ensure thorough testing and preventing mistakes – a possibility when someone may be testing hundreds of appliances one after the other – by prompting the user. Also, data logging in on-board memory and connections to PCs for data storage and manipulation are features. There are various software packages available for this operation if it is desired.

Megger PAT3

This machine offers the required tests in a convenient and easy to use package at a reasonable cost. Housed in a very tough yellow plastic case it has the following specification.

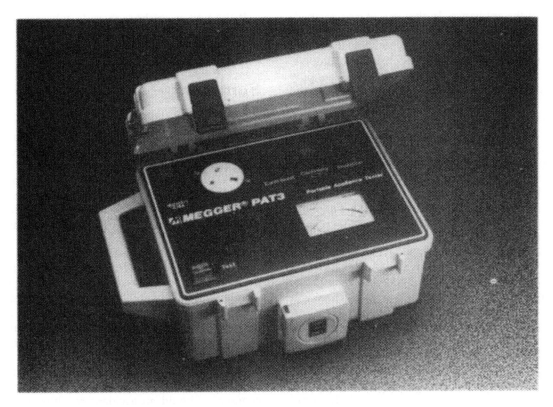

Figure A2.1   *An ideal PAT for a brown goods workshop (Megger Instruments/AVO)*

The initial test for the appliance is a fuse check/continuity test. Naturally, if the fuse were open circuit or the appliance not switched on further tests would be a nonsense. The check is based on full or no meter deflection, indicating continuity or not. With some of our equipment the results may not be as black and white as that. Some less than full scale deflections and decaying readings will be experienced where certain circuits are present across the mains input. This is particularly so with SMPSUs.

A 100 V open-circuit voltage is applied during this test and the short-circuit current is 0.5 mA.

Earth continuity

The earth bond test as required results in a meter deflection indicating from 0 to 0.5 Ω. The pass limit is 0.1 Ω. The 6 V a.c. open-circuit voltage is applied at a nominal 38 A, which at 0.1 Ω equates to 26 A and at 0.5 Ω to 10 A. More sophisticated PATs offer a choice of two test currents, typically a lower 6 A one for appliances with lighter leads such as earthed soldering irons or lamps.

One needs to consider the reason for high current earth tests. Imagine that the earth connection in the plug top or where it bolts to the case of, say, a microwave were damaged and only a couple of strands of wire were actually connecting beyond the PVC sheath. In the event of a heavy fault current flowing to earth these strands could fuse and the earth become inoperative, thus leading to a potentially fatal live case. The earth continuity test will thus show up any deficiency.

Insulation resistance

The single test that most associate with electrical safety testing. This tests the resistance between live and neutral connections and exposed metal parts of the appliance. In the case of non-earthed, Class II appliances an earth return lead is applied to such a metal part – case, aerial socket and ground connection on scart sockets are examples. Here a 600 V nominal, 500 V at 2 MΩ, test supply is applied between live and neutral pins and the earth return. This is one of the benefits of a PAT – as the test unit's mains plug is inserted into the PAT a good connection to both live and neutral pins is ensured. Many pieces of equipment have in the past been damaged by a poor connection, causing the voltage to be applied across live and neutral!

The pass limit on the PAT3 is 2 MΩ. There is a standard for newer units of 4 MΩ.

The correct insulation resistance values for satellite receivers will vary. Where a linear power supply is used and the mains transformer provides isolation, the reading should generally be infinite. Where a switched mode supply is used, there will be a resistor and capacitor between hot ground and chassis ground. The value of this resistor will determine the insulation resistance. Add 1 MΩ or so to its value to gain approximate insulation resistance – e.g. a 10 MΩ resistor will give a unit insulation resistance of around 11 MΩ. In some designs, the value is as low as 5 MΩ. In some more recent designs, the resistor is omitted and thus the insulation resistance is again infinite. Do ensure that you know what you are doing. *Never* assume that an infinite reading is correct – these resistors do go open circuit.

If the unit is earthed, then do take care to not test for too long with a 25 A test current. The lead will begin to get rather warm!

Summary

One needs to remember that all tests should be made when the repair is totally complete and the unit fully assembled. Any test failures must be corrected or the unit removed from use. To reiterate, test results must be recorded. The appliance ought also to be marked. Suitable labels are available from spares suppliers. 'Tested for electrical safety' is supplemented by the date and the initials of the tester.

BS 415: 1990 is the safety standard, laid down by the British Standards Institute, with which a product must comply to gain BEAB approval. Manufacturers ensure at the design stage that their products meet the standard required, and it is the repair technicians' job to ensure that they continue to do so by maintaining the standards detailed above and throughout the book.

It is all too easy to neglect safety in the interests of 'greater efficiency' but this is a gravely false economy. Should an accident resulting in death, fire, injury or other physical damage be caused by a safety defect on a unit repaired and not tested, the engineer who carried out the repair is open to legal prosecution for negligence and also liability for the resulting loss.

Appendix 3

ADDRESSES

Manufacturers (UK where possible, spares departments)

Alba, Harvard House, 14–16 Thames Road, Barking, Essex IG11 0HZ. Tel. (0181) 787 3100, Fax (0181) 377 3110. For spares, see also CPC, WVE and Wizard.

Akai, Haslemere Heathrow Estate, 12 Silver Jubilee Way, Parkway, Hounslow, Middlesex TW4 6NQ. Tel. (0181) 897 6388. Fax (0181) 759 6118.

Amstrad, Brentwood House, 169 Kings Road, Brentwood, Essex CM14 4EF. Tel. (01277) 236111, Fax (01277) 209559. For spares, see CPC and WVE.

Bang & Olufsen (Beosat), Unit 630, Wharfdale Road, Winnersh, Wokingham, Berkshire RG41 5TP. Tel (0118) 927 7804.

Best, Communicado, Flitwick Mill, Greenfield Road, Flitwick, Bedfordshire MK45 5BE. Tel. (01525) 716715. Fax (01525) 716016.

BT, West Side, London Road, Hemel Hempstead, Hertfordshire HP3 9YF.
Note various units are badged Cambridge units.

Bush, See Alba.

Cambridge Industries, Cambridge House, S. Commerce Park, Brunel Road, Teale, Berkshire RG7 4AB. Tel. (01189) 306699. For spares, see SEME.

Channel Master, Glenfield Park, Northrop Avenue, Blackburn, Lancashire BB1 5QF. Tel. (01254) 680444. For distributors, see Longreach Marketing and Eurosat Distribution.

Chapparal, Steve Chilver Trading, Abacus House, Manor Road, Ealing, London W13 0AZ. Tel. (0181) 566 7830.

Comet Group, Service Dept., Unit 5, City Park Industrial Estate, Geldered Road, Leeds, LS12 6DR. Tel. (0113) 231 1024. Fax (0113) 231 1463.

Concentric Satellite Systems, Delta Control, Island Farm Avenue, West Moseley KT8 2UZ. Tel. (0181) 979 9334.

Connexions plc, Unit 3, Travellers Close, Travellers Lane, Welham Green Hertfordshire AL9 7LE. Tel. (01707) 272091. Fax (01707) 269444.

Currys/Dixons/Mastercare Group, 200 The Campus, Maylands Avenue, Hemel Hempstead, Hertfordshire HP2 7TG. For spares: Partmaster. Tel. (01442) 888444. Fax (01442) 888145.

Decca, See Tatung.

Discus, Unit 9, Block 22, Kilspindle Road, Dundee DD2 3JP. Tel. (01382) 833651. Fax (01382) 832418.

Drake, Alston Barry, 32 Broad Street, Ely, Cambridgeshire CB7 4AH. Tel. (01353) 669009.

Echostar, Schuilenbruglaan 5A, 7604BJ Alemelo, The Netherlands. Tel. 0031 546 815122.

Ferguson, Crown Road, Enfield, Middlesex. Tel. (0181) 344 4412. Fax (0181) 344 4452. For spares see also: CPC, CHS, WVE, HRS Electronics, SEME, Wizard.

Fidelity, See Amstrad.

Finlux (badged Cambridge). For spares, see Nokia, CPC, WVE or Akai.

Goodmans, Units 2 & 3, Mitchell Way, Portsmouth, Hampshire PO3 5PR. Tel. (01705) 673763. See also Alba and Comet Group. Some units are Pace clones.

Grundig, Mill Road, Rugby, Warwickshire CV21 1PR. Tel. (01788) 577155. Fax (01788) 562354. For spares, see WVE.

Grundig Satellite Communications (GSC), Unit 10, Lantrisant Business Park, Lantrisant, Mid Glamorgan CF7 8LF. Tel. (01443) 220220.

Hinari, See CPC or Alba.

Hitachi, Hitachi House, Station Road, Hayes, Middlesex UB3 4DR. Tel. (0181) 569 1975. Fax (0181) 569 1441. For spares, see CHS.

IRTE, Fielding Road, Cheshunt, Hertfordshire EN8 9TL. Tel. (01992) 624777.

ITT, see Nokia.

JVC, JVC Business Park, Priestly Way, Staples Corner, London NW2 7BA. Tel. (0181) 450 3282. Fax (0181) 452 2534. For spares, see WVE.

Logik, See Currys.

Luxor, See Nokia and Akai.

Manhattan, brand name used by Eurosat. See distributor list.

Maspro, Electro-Tech, Unit 6, Drury Way Industrial Estate, Laxcon Close, London NW10 0TG. Tel (0181) 451 6766.

Matsui, brand name used by Dixons stores group (inc. Currys). See Partmaster under Currys/Dixons.

Mimtec, 2 Hutton Square, Brucefield Industrial Estate, Livingstone, Scotland EH54 5DD. Tel. (01506) 416262. Fax (01506) 415871.

Mitsubishi, Electric (UK) Ltd., Travellers Lane, Hatfield, Hertfordshire AL10 8XB. Tel. (01707) 276100. Fax (01707) 278859.

NEC, Spares from SEME.

Nokia, For satellite spares, NCS, Bridgemead Close, Westmead, Swindon, Wiltshire SN5 7YG. Accounts only. Non-satellite spares from Akai. Otherwise use CPC or WVE.

NordMende, See Ferguson.

Pace Microtechnology, Victoria Road, Saltaire, Shipley, West Yorkshire BD18 3LF. Tel. (01274) 532000. Spares (01274) 537129. Technical advice (01274) 537122. Returns (01274) 537126. Fax (01274) 537128. See also WVE, HRS Electronics and CPC. Pace WWW site on Internet is at http://www. pace.co.uk.

Palcom, 2a Cleeve Court, Cleeve Road, Leatherhead, Surrey KT22 7NN. Tel. (01372) 360337.

Panasonic, Willoughby Road, Bracknell, Berkshire RG12 8FP. Tel. (01344) 860133. Fax (01344) 861598. Account holders only. Otherwise use SEME. Mainly Pace clones (TUSD100 is Panasonic.)

Philips, PO Box 97, City House, 420/430 London Road, Croydon. Tel. (0181) 686 5414. Fax (0181) 681 0796. Account holders only. Otherwise see CPC, WVE, SEME, CHS, Wizard.

Proline, Brand name used by Comet.

Pye, Brand name used by Philips.

Saba, See Ferguson.

Saisho, Brand name use by Dixons Stores Group. See Currys.

Sakura, Unit 717, Tudor Estate, Abbey Road, Park Royal, London NW10 7UN. Tel. (0181) 961 1600. Fax (0181) 961 2259.

Salora, See Nokia.

Samsung, Unit C, Stafford Park 12, Shropshire TF3 3BJ. Tel. (01952) 207171. Fax (01952) 293459. See also CPC, WVE.

Sanyo, Otterspool Way, Watford, Hertfordshire WD2 8JF. For non-accounts, see CHS.

Sharp, Thorp Road, Newton Heath, Manchester M10 9BE. Tel. (0161) 205 2333. For spares, see WVE.

Sony, PO Box 58, Newbury, Berkshire RG13 4LQ. Tel. (01635) 860000. Fax (01635) 874099. Accounts only. See also CHS, CPC, WVE.

Strong, 49 Berkeley Square, London W1X 5DB. Tel. (0171) 491 7474.

Tatung, Stafford Park 10, Telford, Shopshire TF3 3AB. Tel. (01952) 290111. Fax (01952) 292096. Trade only. See also Wizard.

Technisat, Satellitefernsehprodukte GmBH, Postfach 560, 54541 Daun, Germany. Tel. 0049 6592 712600.

Telefunken, See Ferguson.

Thomson, See Ferguson.

Toshiba, Units 6/7, Admiralty Way, Camberley, Blackwater, Surrey GU15 3DT. Tel. (01276) 694000. Fax (01276) 600521. Trade only. See also CPC, CHS, WVE and SEME. Many units are Pace clones.

Trac, Commerce Way, Skipper Lane, Middlesborough, Cleveland TS6 6UR. Tel. (01642) 468145.

Triax, Saxon Way, Melbourn, nr. Royston, Hertfordshire SG8 6DN. Tel. (01763) 261755. Fax (01763) 262536.

Uniden, See Eurosat Distribution.

Vortec, See Samsung.

Spares/equipment suppliers and distributors

Aerial Techniques, 11 Kent Road, Parkstone, Poole, Dorset BH12 2EH. Tel. (01202) 738232. Fax (01202) 716951. Suppliers of satellite and aerial equipment, specialists in DX-TV.

CHS (Chas Hyde and Son), Prospect House, Barmby Road, Pocklington, Yorkshire. Tel. (01759) 303068. Fax (01759) 303620. Also Viewdata ordering system **CHESS**, Tel. (01759) 306660. Spares and component supplier.

CPC plc, Component House, Faraday Drive, Fulwood, Preston, Lancashire PR2 4PP. Tel. (0172) 654455. Fax (01772) 654466. Spares and component supplier, also satellite dishes and LNBs.

Eurosat Distribution, Oxgate Lane, London NW2 7JA. Tel. (0181) 452 6699. Fax (0181) 452 6777. Local branches also – consult local directory. Satellite and aerial equipment and accessories.

Global Communications, Winterdale Manor, Althorn, Essex CM3 6BX. Tel. (01621) 743440. Fax (01621) 743676. Satellite switches and distribution equipment.

HRS Electronics, Electron House, 100 Great Barr Street, Birmingham B9 4BB. Tel. (0121) 766 6668. Fax (0800) 212179 (freefax). Also Truedata electronic ordering on (0121) 753 0600. Spares and component supplier, also satellite dishes and LNBs.

JW Hardy, 231 Station Road, Stechford, Birmingham B33 8BB. Satellite and aerial equipment and accessories.

Longreach Marketing, 6 Riverside Business Park, Lower Briston Road, Bath, Avon BA2 3DW. Tel. (01225) 444894. Fax (01225) 448676. See also local directory for branches/agents. Satellite and aerial equipment and accessories.

MCES, 15 Lostock Road, Davyhulme, Manchester M41 0SU. Tel. (0161) 746 8037/8. Fax (0161) 746 8136. Tuner, r.f. amplifer and LNB repairers and suppliers.

MTD (Martin Turner Direct), Ty Coch, Llangwm, Usk, Gwent NP5 1HD. Tel. (01292) 650367. Fax (01291) 650702. Satellite and aerial equipment and accessories. Consultancy service.

Protel Distribution, Communication House, 879 High Road, London N12 8QA. Tel. (0181) 445 4441. Satellite equipment and accessories.

Satellite Solutions, Unit 1, Hartburn Close, Crow Lane, Northampton NN3 9UE. Tel. (01604) 787888. Satellite and aerial equipment and accessories.

SEME, Unit 2, Saxby Road Industrial Estate, Melton Mowbray, Leicestershire LE13 1BS. Tel. (01664) 481818. Fax (01664) 63976. Panasonic line: (01280) 823523. Spares and component supplier, also satellite dishes and LNBs.

Wizard, Empress Mill, Empress Street, Manchester M16 9EN. Tel. (0161) 872 5438. Fax (0161) 873 7365. Spares and component supplier.

WVE (Willow Vale Electronics), 11 Arkwright Road, Reading, Berkshire RG2 0LU. Tel. (01189) 876444, Fax (01189) 867188. COPS, Viewdata ordering system (01189) 311969. Also at Enterprise Park, Reliance Street, Newton Heath, Manchester M40 3AL. Tel. (0161) 682 14115. Fax (0161) 682 9031. Spares and component supplier, also satellite dishes and LNBs.

Books and magazines

Satellite Trader, 21st Century Publishing, Pearson Professional, Maple House, 149 Tottenham Court Road, London W1P 9LL. Tel. (0171) 896 2700. Fax (0171) 896 2749. E-mail 21c@pearson-pro.com. ISDN (0171) 383 2232.

Swift Television Publications, 17 Pittsfield, Cricklade, Wiltshire SN6 6AN. Tel. (01793) 750620. Fax (01793) 752399. Suppliers of satellite books and SATMASTER software.

Television, Reed Business Publishing, The Quadrant, Sutton, Surrey SM2 5AS. (Editorial). For subscriptions: PO Box 302, Haywards Heath, W. Sussex RH16 3YY. Tel. (01444) 445566. Fax (01444) 445447. Worldwide monthly engineering and technology magazine including articles on satellite TV technology and servicing.

U-View, 4 South Parade, Bawtry, Doncaster, South Yorkshire DN10 6JH. Tel. (01302) 719997. Fax (01302) 719995. Publishers and suppliers of service manual compilation books.

Trade associations

CAI (Confederation of Aerial Industries), Fulton House Business Centre, Fulton Road, Wembley Park, Middlesex HA9 0TF. Tel. (0181) 902 8998. Fax (0181) 903 8719. The CAI publishes a code of practice for the installation of aerial and satellite systems.

Satellite operators

ASTRA, The Progression Centre, 42 Mark Road, Hemel Hempstead, Hertfordshire HP2 7DW. Tel. (01442) 235540.

EUTELSTAT, 70 rue Balard, 75502 Paris Cedex 15, France. 00 331 53 98 47 47. UK helpline: (0117) 921 0117.

SES, Société Européenne des Satellites: Chateau de Betzdorf, L-6815 Betzdorf, Luxembourg. Tel. 00 352 7107251.

INDEX

Printed and bound by CPI Group (UK) Ltd, Croydon, CR0 4YY

03/10/2024

01040335-0015